The Easiest Plants to Grow

315

Easy Does It

Having a beautiful garden can be easy. All you have to do is choose the right plants for the right places. The easy garden can include trees, shrubs, and flowers, even herbs and vegetables. This book will tell you how to have a successful garden—the easy way.

An easy garden is one whose plants do most of the work for you. Although no garden can endure months of total neglect, an easy garden is not demanding. Its care does not require a degree in horticulture or a set of garden encyclopedias. It does not need to be watered, pruned, or sprayed every day. The easy garden succeeds whether or not you think you have a green thumb.

The only requirement is that you find plants that suit your climate, your garden, and your needs. This book includes lists of trees, shrubs, flowers, ground covers, vegetables, herbs, and houseplants to help you in every situation.

This collection of dwarf shrubs includes pines, arborvitae, and juniper. All are easy to grow and require little care.

PLANT ADAPTATION

Success in the easy garden depends upon finding plants that suit you and your garden, not changing your garden to suit the plants. No plant can grow everywhere. Perhaps the dandelion comes closest, but gardeners want plants that are less aggressive and more attractive than dandelions and other weeds. All good garden plants have certain demands, especially for temperature, light, and moisture. If the climate and soil conditions of your garden are similar to those of the area where a plant originated, the plant will probably thrive for you. Northern plants do well in cool gardens; tropical plants in warm gardens; desert plants in dry, sunny gardens; bog plants in wet gardens; and so forth.

Climatic Zones

The biggest single factor that determines what plants will be easy for you to grow is cold hardiness—a plant's ability to survive the winters where you live. Trees, shrubs, and perennial flowers that are not hardy enough may die during the first winter, or they may live a year or two but succumb to an especially cold winter.

The easiest way to figure out what will survive is to see what your neighbors are growing. The next is to find out what climatic zone you live in. Identify your climatic zone on the map below. The zones of this map indicate the average minimum temperatures everywhere in the country. When you are choosing trees, shrubs, and other perennial plants that must survive the winter, check what climatic zone they can grow in. Generally, you can grow anything for your zone or any zone with a lower number. Lower climatic zone numbers indicate colder winters. For example, winters in Zone 5 are about 10° F colder than winters in Zone 6.

Plants native to your area are likely to be some of the most dependable things you can grow. Many indigenous plants are given in the lists that follow. But there are also many nonnative plants from climate and growing conditions similar to yours that will do perfectly well for you.

Microclimates

Of course, climate is only one of the factors that helps make your garden unique. Other factors that determine your microclimate—that

Climate Zone Map

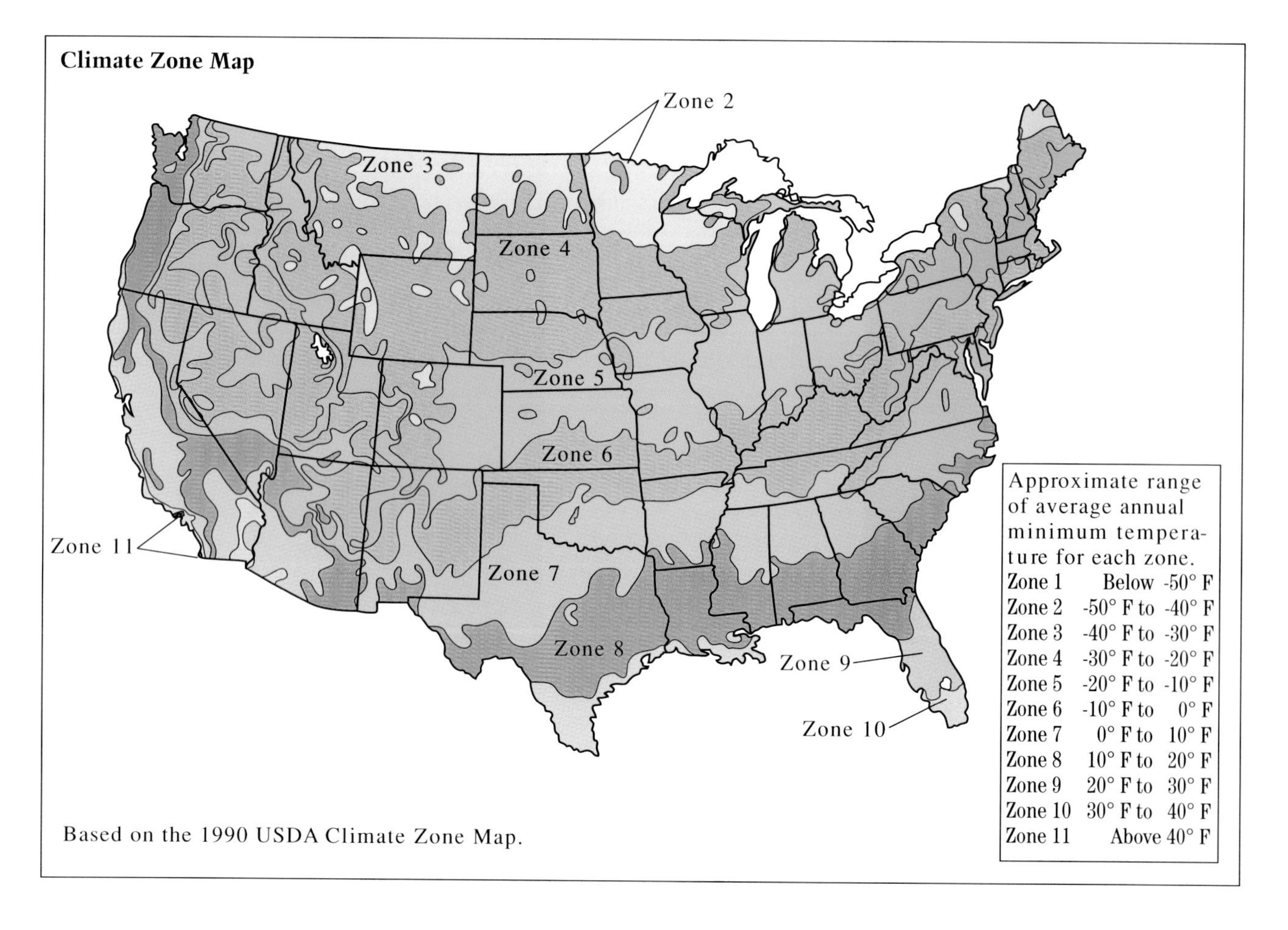

Based on the 1990 USDA Climate Zone Map.

is, the climate specific to your garden—include the degree of slope and the patterns of sun and shade. You can grow a garden in any microclimate, no matter how limiting you think it is. You don't have to change your topsoil or install expensive drainage systems. There are plants that thrive in dry soil in sun, others that do well in constant shade, some that can endure a lot of wind or dampness or even air pollution.

Having an easy garden means that you'll probably have to do without fussy plants—those that need a lot of pruning, watering, mulching, and spraying. But in the compromise, you are likely to find many plants just as beautiful as the ones that take a lot of attention.

WHAT MAKES A PLANT EASY?

To be considered easy, a plant needs to be pest resistant, unaggressive and noninvasive, and able to prosper without regular attention.

Pest Resistance

An easy plant is one that will stay healthy without frequent inspection or constant spraying. Certain plants that would otherwise be perfectly adapted to your garden, and may even be native to the area, are susceptible to insects or diseases that make them less than ideal in an easy garden. This does not mean that your garden will be free of insects. They are part of any garden. Some caterpillars and butterflies are decorative or beneficial and so are welcome parts of a healthy garden. Others are pests that trouble all kinds of plants but are fairly easy to get rid of. If you do find garden pests or disease problems, ask for help from a garden store or an extension agent.

Good Manners

The dandelion would be an ideal candidate for the easy garden if it were not quite so successful at growing and reproducing. Unfortunately, there are many plants such as dandelions that take care of themselves entirely too well. You may have expected to find some of these plants in this book, but they aren't here because they are simply too aggressive. What may seem at first to be an easy plant ends up demanding cutting back and rooting out. Then you discover that you have what you did not want—a difficult garden.

Many of these invasive or aggressive plants, such as dandelions, are called weeds. Many others are perfectly acceptable in some places in the garden but too pushy in others. They may be welcome where winters are very cold or conditions are dry and harsh, or in a spot surrounded by concrete or by lawn, where mowing keeps them confined. In these situations, their toughness allows them to survive, but the climate, mowing, or paving keeps them from getting out of hand. If they were grown in fertile soil in a temperate climate, or grown next to more reserved plants, they would quickly spread beyond your desires, and maybe your neighbors' too.

On the other hand, some plants, such as common lilac, mint, and snow-in-summer, are so ornamental or useful in the right places that they have been included in this book, even though they are too pushy for some situations.

Other plants are messy. The worst of these are trees with brittle or weak wood that breaks in strong winds or under the weight of ice or snow. These trees can actually pose a danger to life and property, so they have not been included in this book. Messy but less worrisome are trees that drop flowers, fruit, and leaves. In an informal setting, you may not mind what your

Azaleas and hostas thrive in this shady microclimate. The azaleas particularly like the good drainage afforded by the sloping location.

tree drops. But if you really dislike raking leaves or if you want a tree over a patio or pathway where fallen fruit would be unsightly as well as slippery, something coniferous or small leafed might be the best choice. Otherwise, cleanup will demand your time.

Self-sufficiency

Plants have greater and lesser degrees of self-sufficiency. Many of the most beautiful garden plants need plenty of attention. If you want to grow flowering shrubs, for example, chances are you will have to do some pruning after flowering to encourage next year's bloom. If you want a formal hedge rather than an informal one, it will need some pruning. If you want to grow vegetables, you will probably have to do some summer watering and fertilizing. All perennial flowers eventually need dividing or renewing. Plants in containers or window boxes demand more attention than plants in the ground, if for no other reason than they dry out faster.

Here is one way to deal with messy trees. This low-growing camellia has been planted as a ground cover under an ornamental pear that drops leaves all winter in the mild climate of this location. The ground cover is deep enough to absorb the leaves; they filter down into the foliage, forming a natural mulch, and never need raking.

How much time you want to spend in the garden is up to you. If you create an informal, relaxed-looking garden, you will have an easier time than if you want everything to look perfectly groomed.

If you choose to grow plants that demand some attention, here are some of the factors you should be prepared to deal with. Some plants have very particular soil requirements. If you aren't blessed with the right soil, you will have to change your soil to suit them, and from then on you must pay attention to the soil to make sure it gives the plants what they need.

Some plants are very sensitive to drying out; others cannot tolerate much water around their roots. If you have exactly the right conditions, these plants might be easy; otherwise you will be tied to daily watering or the installation of an irrigation system or a drainage system.

Many plants need pruning at least once a year. Almost all fruit trees and berries have this requirement, which is why they cannot be considered truly easy plants and are not included in this book. Fruits and berries are also among the plants most troubled by pests and diseases.

Winter protection can also demand your time and attention. Perennial plants too tender for your area need to be brought indoors in fall or protected outdoors. The less suited the plant is to your climate, the more protection it will demand. Northern fanciers of hybrid tea roses are willing to go to great lengths to insulate their precious plants, but if you want an easy garden, you should choose roses that can survive without any special care.

THE PLANT LISTS

The plants in this book are listed by their most important garden features—for instance, size of plant and flower color—then by their common name followed by their Latin, or botanical, name.

The botanical name of a plant is a more dependable means of identification than its common name. For example, the common name of *Hibiscus syriacus* is rose-of-Sharon. It is a beautiful shrub described in the chapter on trees and shrubs. But there is another plant whose common name is also rose-of-Sharon. *Hypericum calycinum* is a ground cover with big yellow flowers. The two plants are very different. On paper, only their botanical names or a picture can clear up the confusion.

You can be certain of getting the plant you want only if you know its entire botanical name. The first two words in the name are Latin. The first word, which is capitalized, is the genus—the group of plants yours belongs to. *Hibiscus syriacus* belongs to the genus *Hibiscus*. The second word is the species name—in this case, *syriacus*. The species name narrows your choice from some sort of hibiscus to just one type of hibiscus—the shrubby one with big pink flowers and a certain set of other characteristics. To anyone who knows a little Latin, the species name can be helpful. The word *syriacus* means that the plant comes from Syria, suggesting that this is a plant that likes warm, dry conditions like those in much of the eastern Mediterranean.

Many plant names are further clarified by a third word. If it's in italics, it's the variety. If it's capitalized and in single quotes, such as 'All That Jazz', it's a cultivar—in this case, a type of rose. Cultivar means a cultivated variety. A cultivar is the work of human beings. There may be many, many cultivar names for just one species, each with different characteristics, and you'll need to know the cultivar you want when you look for your plant. Instead of the word *cultivar*, however, most gardeners simply use the word *variety* to mean the same thing.

USING THIS BOOK

This book is arranged by plant category, such as Large Shade Tree and Low Ground Cover for Shade. One or more plants are described in each category. In each case, the plants described perform that landscape function well and are the plants most likely to be satisfactory in that position.

The plants listed in this book are resistant to pests and diseases and tolerant of difficult conditions, such as poor soil or harsh weather. They are the "easiest" plants available in nurseries. Some—perhaps even most—are common for just that reason: they are the most satisfactory plants that do the job. Others are less well known but will probably become more common in the future as people discover their fine traits.

The best way to select a plant is to define what you'd like it to do for you. The more traits you can list, the easier it will be to select a plant. For example, you might want a shrub to fill a corner of your backyard. If you decide that you'd like a shrub high enough to hide a fence, and one that offers a display of fragrant flowers for a few weeks in spring, you might select 'Excel' lilac, a pink version of the common lilac that is hardier and more disease resistant than the common form.

Before buying a tree or shrub, it is wise to search out an example of it in the landscape so you can see what the mature plant will look like. A local arboretum probably has labeled specimens on display in a natural setting. A common error in purchasing plants is to forget how large they will grow in the garden. Some of those cute little plants in the nursery will grow into giant trees that can crowd out other plants or push into the eaves of your house if planted too close to it. The big ones should be given lots of room.

You can't go wrong if you define carefully what you want, then select from the plants listed in this book. Your landscape will not only be beautiful, it will be satisfactory and easy to care for as well.

Delphiniums, although beautiful, are difficult plants to grow well without ideal conditions. They require staking if any wind is present, the soil they are planted in must be specially prepared, they need careful watering, and they are prone to insect and disease problems, which make spraying a necessity.

Starting Out

Planning and preparation are the keys to success in an easy garden. Choosing the right plants and then planting properly set the stage for a garden that practically takes care of itself.

A successful garden doesn't have to mean struggling with the site, soil, and plants. It does mean avoiding fantasies about growing certain plants, no matter how impractical. If you want an easy garden, you need to accept what your garden has to offer—sun; shade; cold winters; hot summers; soil that is wet or dry, acid or alkaline—and choose your plants accordingly. Working with your garden instead of against it makes the entire process easy and fun. It also broadens your scope, encouraging you to think about some interesting, beautiful plants you may not have considered before.

After reading this book, spend some time studying your garden and thinking about it. Notice where the patterns of sun and shade fall, where it's windy, and where the water puddles in rainy weather. Then select plants from this book that match each of those conditions, and that you like. If the plants you select are hardy where you live, they won't require any particular winter protection or care. If you live in the North, or in cold mountain country, select from the plants noted as being especially hardy.

The secret to an easy garden is to plant a garden that is easy. That sounds simplistic, but it is true. Find out what makes a garden easy, plan one that meets those specifications and also satisfies you, and the garden will reward you by being beautiful without extra care on your part.

PLANNING THE EASY GARDEN

You can guarantee garden success if you put time and thought into the planning process. Select easy-care plants and design the best garden for your site and preferences.

Working With What's There

Chances are that some plants are already growing in your garden. Some of them may be difficult to take care of. They may be the pampered pets of the former owner of your house. They may be old or diseased, or perhaps you just don't like their looks.

Since you are aiming for an easy garden, this is the time to get rid of plants that do not fit your new plans. You may have to hire a tree removal specialist to take away trees and shrubs that are more trouble than they are worth. You may be able to dig up and dispose of smaller plants yourself. Check with your neighbors or a local horticultural society to find out if anyone might welcome the plants you don't want.

Most likely, you will want to keep some of the plants that are already in your garden, especially those that take care of themselves. You can work around them, avoiding their roots and allowing for the eventual size of any new plants by spacing them carefully.

If your yard has not yet been landscaped, you'll be able to start from scratch. If you have nothing but lawn and want something more interesting, begin by figuring out where you would like plants to grow, then get rid of weeds and lawn in those areas.

Designing Your Garden

It is probably obvious that the smaller your garden, the less trouble it is likely to be. Even if you own a lot of property, it is wise to start small and add gradually to your garden. That way, you will never be overwhelmed by too much work.

Trees, shrubs, and low ground covers are the easiest plants to grow. Flowers and vegetables take more time and attention. Lawns, with their need for regular mowing, probably take the most time. Figure out how much time you want to spend in the garden before you put in beds for flowers or vegetables, and make these initial beds small. You can always make them bigger later on.

In general, the more formal your landscape, the more trouble it is to care for. On the other hand, the more it looks like the natural landscape in your area, the more self-sufficient it will be. A formal landscape requires lots of shearing and pruning to keep hedges and

Although neat and attractive, this landscape requires no routine pruning. The hedges are informal and the shrubs have been selected for their shape and size so they don't need additional trimming.

shrubs looking tidy, whereas plants in an informal landscape can be allowed to follow their natural growth patterns without intervention from the gardener.

Trees and shrubs are likely to be the largest plants in your landscape, so plan for them first. In planning where to put them, make allowance for their eventual height and width. Their shade might be beneficial or detrimental. A patio that is shaded in summer can be lovely, and shade against the house can help keep it cool. On the other hand, shade all day on your perennial bed will drastically limit what you can grow.

Trees that become too large for their sites are difficult and expensive to remove. They look out of balance in their surroundings; worse, they will probably shade or crowd the gardens of your neighbors. Plan carefully when situating any large tree.

Preparing the Beds

You should remove any existing sod and weeds and the larger loose rocks in areas where you would like to put flower or vegetable beds. Shake the soil off the sod before removing it. The bed's surface will be a little lower than the surrounding area after the sod is removed, so build it up again with purchased topsoil or with compost. The surface of the soil in your planting bed should be a couple of inches higher than the surrounding soil or lawn. Make the bed large enough so that the roots of the new plants can spread without competing with those of other plants.

Most of the plants in the following lists will grow in any ordinary soil, even if it contains a lot of sand or clay. You can increase your soil's humus content by digging in compost or good topsoil, but otherwise you can leave it as it is.

Obtaining Plants

Reputable local nurseries and greenhouses are usually good sources of easy-care plants. You may have to sort through what is fashionable rather than what is easy. Although most of the plants recommended in this book are popular and easy to locate, you may have to search a little for some of them. Mail-order nurseries will ship at planting time. Some of these companies carry a little of everything; others specialize in trees, shrubs, roses, perennials, herbs, houseplants, or vegetable seeds and transplants. Advertisements in gardening magazines give the addresses and catalog prices of these nurseries.

INTO THE GARDEN

No matter how easy you want your garden to be, it must be planted properly. This is not the time to cut corners. The month after transplanting or sowing seed is also a very vulnerable time for new plants or seeds. Perennials need a longer time of care at the start than annuals. Only when plants become established and are actively growing can you give less attention to the garden.

Transplanting

Trees, shrubs, and perennials should be planted within a day of your bringing them home. This is especially important with anything purchased bare root or balled and burlapped. Potted plants can wait longer, but they should be given a thorough watering as soon as possible, placed in partial shade, and not allowed to dry out. Before you buy or order plants, you should decide where the newcomers will go so that they do not spend a lot of time sitting around.

Hot, windy weather puts the most stress on plants. If the weather is sunny, wait until late in the day for planting, so the plants can have a night's rest in their new homes. A cool, misty,

Plant trees and shrubs in holes that are twice the diameter of the root ball. This surrounds the new tree with soft, dug-up soil that its roots can easily penetrate.

This newly planted tree is being staked, with the stakes at right angles to the prevailing wind. The next step is to cut off the stakes just above the ties so the tree doesn't rub on their ends.

even rainy day is also an excellent time for planting.

Dig a hole that is a little bigger than the root system of the plant, and the same depth as the rootball. Set aside the topsoil and discard any rocks. The plant should grow at the same depth it grew at the nursery or in its pot. Keep the stem upright while planting. You will need some help with a tree—one person to hold the plant while the other fills in around it with soil.

If there is danger of a newly planted tree blowing over, or if it is so spindly that it won't stand up by itself, you will need to stake it. Drive a stout stake into undisturbed soil on the windward side of the plant. Tie the trunk to the stake as low as possible and still have it support the tree. The more the tree is allowed to move in the wind, the sooner it will be able to stand on its own.

Tie stake and tree together with a strip of cloth. Cut off the top of the stake just above the tie. Stakes can be purchased at garden supply stores, or you can use a recycled stake such as a shovel handle. The stake should remain in place for the first year.

If the location is particularly windy, use two stakes at right angles to the wind direction, tying the tree between them with loops of cloth.

Fill in around the roots of the tree with soil, making sure you do not leave any air pockets. Add compost or purchased topsoil if you do not have enough garden soil. Make a little mounded circle around the perimeter of the filled hole to form a watering basin. Fill the basin with water until the backfill soil is saturated and muddy. Wiggle the plant to free any air bubbles. If the weather is dry, leave the basin in place for a few weeks to make watering easier. Until the plant is established, water frequently to keep the soil around the roots from drying out.

If a tree or shrub will be surrounded by lawn, leave a circle of clear ground around the stem so that you can easily mow near it. Keep the circle free of grass and weeds with a layer of landscape fabric (see opposite page) and cover it with bark chips or small stones so the fabric won't deteriorate in the sun.

Sowing

A few easy-care plants can be grown from seed rather than purchased transplants. The larger the seed, the deeper it needs to be set. The rule of thumb is to plant a seed about twice as deep as the seed's width; for example, a sunflower seed that is a half inch wide should be planted an inch deep in the soil.

Soil that is to be seeded must be clear of weeds. Water the soil before sowing your seeds. You may want to mark the positions of individual seeds or rows or patches of seeds with sticks or markers, so that when the seeds sprout you can identify them and not mistake them for weeds. The soil should be kept moist until the seeds sprout. In very windy or dry gardens, you can make this job easier by mulching the seeded area with a light covering of branches, fern fronds, or cut grasses. After the seedlings sprout, water them whenever the surface of the soil dries out.

Weeding and Mulching

Few people enjoy weeding. Fortunately, the easy garden will eventually have very little open ground where weeds can sprout. But when plants are young and there is empty space around them, they can be choked out by weeds. For the first year, you should check your garden every week or two for weeds. As your plants mature, they will fill in and shade the ground, discouraging the growth of less desirable plants.

Mulching lessens the need for weeding. You can cut down on weeding chores by filling open areas with ground covers or a material that will keep weeds from sprouting. A layer of landscape fabric—a special synthetic cloth that allows water to pass through to plant roots but keeps weed seeds from germinating—can be spread on the ground either before or after trees and shrubs are planted. If the fabric is already in place, you can cut holes in it for planting. Over the fabric, place a layer of some material deep enough to keep the sun from striking the fabric, causing it to deteriorate. You can use bark chips or pebbles or even a thick layer of fallen leaves or grass clippings. If you use an organic material such as plant clippings or leaves, be prepared to add more as it breaks down, so the fabric is always covered.

You can substitute several layers of newspaper for landscape fabric. Plain newsprint is better than shiny, color-coated paper. The newspaper should be entirely covered to make the garden look attractive and to keep the paper from drying, tearing, and blowing in the wind. The paper will compost into soil in about a year and may need to be replaced.

An organic or a stone mulch will also control weeds without newspaper or fabric underneath, but it should be about 4 inches deep. An organic mulch will decompose, and more should be added to keep the depth constant.

Watering

Choosing plants suited to your climate keeps watering chores to a minimum. For dry gardens, many plants are surprisingly resilient even when it doesn't rain for weeks.

But all new transplants need water. For the first year, water shrubs and trees deeply once a week if the weather is dry. Perennial flowers need to be watered every few days for the first month, annuals every day for a week. You don't need to water the entire garden, just the new plants. Some plants will signal their need for water by wilting. Wilting during a hot, dry day is not necessarily a sign that watering is needed, but if plants are still wilted in the morning, they must be deeply watered. The water needs of some plants, such as coniferous trees, may be more difficult to discern. Woody plants that are not sufficiently watered may look alive at the end of summer but may not be strong enough to survive their first winter. Play it safe and water recent transplants until they signal their success by resuming growth.

A mulch of wood chips controls weeds in this first-year perennial bed. In a few years, the bed will fill in with flowers, and the mulch will no longer be necessary.

You can save time and water by watering the ground under the plants rather than the foliage. If you water directly on the ground, by hose or a drip irrigation system, you can water late in the day, when evaporation rates are lowest and the soil will stay moist the longest. If you have a sprinkler system, however, you should water in the morning, so the leaves can dry during the day. Leaves that remain wet overnight can invite mildew.

Fertilizing

An easy garden seldom needs fertilizing. In fact, fertilizing can encourage rapid, lush growth, which means more weeding, more mowing, more pruning, and often more diseases and pests. Fertilizing can actually reduce flowering and fruiting. In the easy garden, plants are usually content with the existing soil. Plants should be fed enough to keep their lower leaves from turning yellow, a sign of nutrient deficiency. In general, one light sprinkling of fertilizer over the entire garden in midspring, when growth is most rapid, is enough for easy-care plants. Vegetables and annual flowers, which should continue rapid growth, should be fed once or twice more as long as they are growing rapidly. Fertilizing should be stopped by midsummer so plants will have time to prepare for winter, a process called hardening off.

The Easiest Trees, Shrubs, and Vines

Trees, shrubs, and vines give your garden structure and stability. They are the tallest living elements in it, and will probably last the longest, so choose carefully.

In choosing a tree or shrub for your garden, remember that they can create a considerable amount of shade. Before you plant, look down and around, especially on the north side, to see where the shade will fall. Not only will the shade affect the atmosphere of the garden, it will also limit the choice of plants that you can grow in these areas. While you're at it, look up too; branches may eventually snag overhead wires or block views from upper windows.

Trees and shrubs will widen as well, both above ground and below, so space should be left between them and walls, fences, or other plants. If you keep in mind that the tree, shrub, or vine you will eventually have will be larger than what you plant, you will be happy with your long-term investment in beauty and shelter.

If you are planting a hedge, remember that formal hedges require pruning at least once a year. For a really easy garden, grow an informal hedge, one that is left unpruned and so has an irregular shape.

For privacy, a hedge should be at least 6 feet tall. Shorter hedges can be used to define garden beds. For a hedge 4 to 6 feet tall, space the plants 18 inches apart. For a shorter hedge, set the plants closer together; for a tall hedge or windbreak, place them 2 to 3 feet apart.

To select a tree, shrub, or vine, decide what job you'd like it to do and search the chapter for the one that meets that need.

A crab apple gives a spectacular spring display. These deservedly popular flowering trees can be easy to grow if a disease-resistant variety is selected. See page 21 for some of the best.

TREES, BROADLEAF

Large Shade Tree

Red maple
Acer rubrum

To many people, a maple tree is the ultimate shade tree. The distinctive foliage is beautiful, from its unfolding in spring through its provision of dappled shade in summer to its eye-catching colors in fall. Blazing red fall color is one hallmark of the most popular shade tree in the United States, the red maple. The new growth in spring is also bright red. The tree, a North American native, is hardy from Zone 3 south, though it does not do well too far south, in desert areas, or anywhere with extreme heat or drought.

The red maple grows fairly quickly, as fast as 2 feet a year to a mature height of about 50 feet in most garden situations. The color development of the fall foliage of red maple, like that of other maples, is somewhat dependent on weather and is especially bright when fall brings dry days and frosty nights. The fall color looks most spectacular against a backdrop of pines or other conifers.

Red maples prefer moist, acid soil, but tolerate a wide range of conditions.

Large Oak

Red oak
Quercus rubra

Oaks are majestic trees, slow to grow but so long-lived that planting one means beginning a legacy of beauty and shade. The red oak, named for the color of the new leaves and leafstalks in spring and the color of the foliage in fall, has a high branching habit and wide crown that make it a beautiful boulevard tree. The red oak grows 40 to 75 feet tall in gardens but can be much taller in the wild. It begins to flower at 20 to 30 years old. This is not a tree for small places or pressing schedules, but where it can grow to maturity, the large-topped crown gives excellent shade. The inch-long acorns are a delight for children.

Left: Red maples are favorites in northern states, particularly prized for their striking fall colors.
Right: Like most oaks, red oak supports a host of insects and minor diseases, but seems to tolerate them all without difficulty. Treatment is rarely needed.

This American species, sometimes called *Q. borealis* var. *maxima,* is easier to transplant than most oaks. It is hardy to Zone 3 but is very sensitive to drought and prefers plenty of water. It does best in deep soil.

Large Elm

Zelkova
***Zelkova serrata* 'Village Green'**

Over most of North America, Dutch elm disease has wiped out or endangered the beautiful American elm, bringing into the spotlight some lesser known but disease-resistant relatives. This Asian member of the elm family, sometimes called the Caucasian elm, has many of its cousin's best features: large size, high-quality hardwood, dependable shade, and a wide crown. Like the American elm, zelkova can grow to more than 75 feet tall, so it is suitable only for rural gardens or large estates. The leaves are elmlike but narrower, and smooth above but hairy on the veins underneath. In fall, they turn rich orange-brown or golden.

Zelkova does best in deep soil. It needs consistent watering when young but in a few years becomes fairly drought resistant. The cultivar 'Village Green' is superior to the others in disease resistance, speed of growth (up to 3 feet a year), and fall color—a deep wine red. Unfortunately, zelkova is far less hardy than the American elm, only to Zone 6.

Large Flowering Broadleaf Evergreen Tree

Southern magnolia
***Magnolia grandiflora* 'Gloriosa'**

No tree suggests the aristocracy and romance of the South like the southern magnolia, a tree that dominates gardens from Zones 6 to 9. In places where it thrives, it is the classic broadleaf evergreen, forming an elegant pyramid as tall as 80 feet. A garden need consist of little else. The glossy leaves are as long as 10 inches, the flowers equally wide. They are cup shaped and fragrant, creamy white on top and often pink to deep purple on the underside. They appear in spring, beginning when the tree is about 15 years old.

Zelkova 'Village Green', which can grow to magnificent size, is resistant to Dutch elm disease.

The southern magnolia demands space, because it can spread as wide as 50 feet. Shelter it from strong winds. Give it deep, rich, moist soil, and do not crowd it or plant directly underneath it, because it has a network of surface roots. It drops a fair amount of litter, but only the most fastidious gardener would exclude it for that reason alone. Fallen leaves and flowers are small failings in a tree that is otherwise virtually problem free.

Newly planted southern magnolias require faithful watering if summers are dry, but once established they are quite self-reliant.

Large Broadleaf Evergreen Tree

Southern live oak
Quercus virginiana

So massive that it dominates even a large garden, the southern live oak grows as tall as 50 to 70 feet and even wider. If you have a place big enough and warm enough for it—Zone 7 or farther south—it is the most beautiful of the evergreen oaks and is also almost problem free. It will likely outlive you by a century or two.

Top left: Southern live oaks bring to mind antebellum mansions and Spanish moss, but they can be just as dramatic in a suburban yard if it has room for their massive stature.
Top right: A Russian olive, with its unusual gray leaves, stands out in a mix of darker green foliage. It is especially attractive planted with dark-colored conifers.
Bottom right: Olive trees are so tough that they can be transplanted even when mature. In California, where olives are raised commercially, trees from old orchards are sometimes relocated to urban landscapes. Varieties that bear fruit can be messy, however. Don't plant them over patios or sidewalks.

This North American native will put up with dry or moist conditions and almost every type of soil. It has waxy, broad leaves that remain attractive all winter. The old leaves drop in spring, revealing the new crop. The acorns attract birds and squirrels. In the arid Southwest, look for the cultivar 'Heritage'.

Medium Tree for Dry Soil in the North

Russian olive
Elaeagnus angustifolia

The Russian olive, or oleaster, is a dependable tree for areas with cold winters (to Zone 4) and poor, dry soils. It can take saltier soils than most trees. It has become a favorite garden tree for the North because it grows to only about 20 feet tall and the foliage is a pleasant gray-green that looks attractive in many positions in the garden, especially in contrast with the reddish or dark green foliage of other trees and shrubs. The flowers, which are very small but fragrant, are followed by small, edible fruit. The Russian olive may assume a shrubby form unless pruned to a single stem.

Russian olive should not be grown in areas where it is susceptible to diseases. They are most problematic in moist soils and in the southern parts of its range—Zones 6 and 7. The worst disease, especially prevalent in the Midwest, is the fungus called verticillium wilt.

Left: Chinese pistache is one of the few trees easy to grow in hot, dry climates that also produce good fall color. Right: 'Liset' crab apples blaze in spring glory against dark evergreens.

Medium Tree for Dry Soil in the South

Olive
Olea europaea

The olive is an effective ornamental for gardens where summers are very hot and dry and winters are no colder than 10° F—Zone 8 or warmer. Its smallish leaves, pale green above and whitish underneath, are an attractive contrast near dark-colored walls or in front of shrubs or trees of darker green. Olive leaves can virtually stop transpiring when the weather is dry, so the tree needs no watering.

Olive grows slowly to a rounded, bushy shape as tall as 30 feet. Side branches form trunks, which eventually become gnarled, giving older trees a sculptural quality. The tree can be pruned to a single trunk and trained to have a pleasing, open shape. It is a good idea to remove any suckers that grow from the base.

Medium Tree for Fall Color in the South

Chinese pistache
Pistacia chinensis

Fall color can be provided to southerly gardens that may not otherwise have it by planting the Chinese pistache. This tree, which is happiest where summers are long and hot, even in the desert, turns brilliant orange, scarlet, or crimson in fall, bringing a touch of the north in Zones 7 to 9. The foliage is 9 inches long and finely divided into multiple segments, resembling that of sumac.

Growth starts slowly, so the Chinese pistache should be staked for as long as 3 years. It can eventually become 40 feet tall and almost as wide. New growth in late spring is reddish. The lower limbs can be pruned off if you want a high canopy.

This tree tolerates all kinds of conditions, including the poorest of dry soil, but it does best in moist, well-drained soil, especially beside a stream or lake.

Medium Tree for Spring Flowers

Crab apple
***Malus* species**

Many crab apples make beautiful garden trees. Most are relatively small, 30 feet or less, with lovely white, pink, or red spring blossoms, sometimes double, followed by edible fruit, which may be red, green, or yellow. There are 20 to 30 species and more than 500 cultivars. 'Adams' is an excellent pink variety; 'Donald

Wyman' is one of the best whites. 'Autumn Glory' holds its fruit for winter interest.

Crab apples are especially valued as dependable spring flowering trees for areas with very cold winters. Many are hardy to Zone 2, and all require a period of chilling for best flowering. For southern gardens, choose the white-flowered 'Callaway'.

Crab apples will do well in almost any soil, but like apples they are healthiest in heavy loam. They should be given full sun. Crab apples are excellent specimen trees or patio trees for city gardens or the back of a perennial border in sun.

Medium Maple for the North

Amur maple
***Acer tataricum* ssp. *ginnala* 'Flame'**

The Amur maple, up to 25 feet tall at maturity, is tough and problem free in almost all areas except those with very hot summers. Sometimes grown as a substitute for Japanese maple in gardens too cold for that species, Amur maple is appreciated for having the first foliage to turn color in fall. Fall color is best in full sun, although the tree will put up with some shade. The leaves turn scarlet or orange-red and remain on the tree for several weeks. The cultivar 'Flame' is especially colorful.

Amur maple will grow in almost any soil, though it does best in moist and well-drained soil. It generally forms a large, spreading shrub unless it is pruned to a single stem. It is hardy in Zones 2 to 8. It can be grown in a large container anywhere except in its northernmost range—Zones 2 to 4—where container-grown trees may suffer winter damage. It appreciates cool climates and tends to be less healthy in Zone 8. Unlike most maples, it has fragrant flowers, which are small and yellowish white.

This Bradford pear is just beginning its spectacular fall display. In mild climates, the color is not so dramatic.

Medium Tree With Spring Flowers

Bradford pear
***Pyrus calleryana* 'Bradford'**

When it arrived on the North American market in 1963, the Bradford pear, or callery, a Chinese relative of the edible pear, quickly became one of America's most popular spring flowering, midsized trees for all kinds of places, from city front yards to country perennial borders. It is a handsome shade tree, about 25 feet tall, appreciated for richly colored fall foliage and its profuse flowering in spring. The white blossoms are so dense that for a couple of weeks the tree looks like a cloud.

The leaves are glossy and dark green in summer. Their fall color is a beautiful scarlet, purple, or even yellow, depending in part on fall weather. Color is most vivid after an early frost. This pear is resistant to fire blight, a troublesome disease of edible pears.

The Bradford pear needs full sun. It puts up with almost any soil except one that is always wet. It is hardy to Zone 5.

Medium Tree With Yellow Flowers

Goldenrain tree
Koelreuteria paniculata

The goldenrain tree, or pride of China, is named for the panicles of beautiful ½-inch yellow flowers that appear in summer. The flowers make the tree a showy specimen in a lawn, behind a border of shrubs, or shading a patio. It is almost trouble free in areas as cold as Zone 5.

The bright-yellow flowers of the goldenrain tree are followed by seed pods that look like small Japanese lanterns. The pods change from green to yellow and brown as the season progresses, providing interest into fall.

The handsome leaves, up to 18 inches long, unfold purplish red late in spring. They are dark green in summer, then turn golden and drop fairly early in autumn. Autumn color is best during a dry fall with cold nights.

The goldenrain tree dominates a large space because it grows about 40 feet tall and wide, and the dense shade underneath permits the growth of little else. It tolerates almost all soil conditions, though it should be given full sun.

Medium Tree With Decorative Fruit

Pagoda dogwood
Cornus alternifolia

Dogwoods are generally easy trees to grow, and this one has the advantages of greater disease resistance and greater winter hardiness than its coastal competitor, *C. florida.* Hardy enough for Zone 3 and healthiest in cool areas, this

In summer, it is easy to see the layered effect of the branches of the pagoda dogwood, which gives rise to its name.

North American native is sometimes called blue dogwood, for its blue fruit. The name pagoda dogwood is derived from its tiers of branches that are almost parallel to the ground, suggesting its decorative value in the winter garden. Its winter beauty is further enhanced by the purplish cast of the stems. The large, yellowish white spring flowers are delightfully fragrant.

C. alternifolia will take the shape of a shrub unless pruned into a tree form, but it is otherwise easy, provided it grows in partial shade and acid soil. It grows 8 to 25 feet tall and somewhat wider, with the smaller mature height in the northernmost part of its range.

Small Tree With Showy Foliage for Humid Areas

Japanese maples are available in hundreds of varieties. Those with red or purple leaves are beautiful all summer; those with green leaves have better fall color.

Japanese maple
Acer palmatum

Like most maples, Japanese maples virtually take care of themselves after planting, but the catch with these, which are among the most graceful of all garden trees, is that they are not particularly winter hardy or heat tolerant. A few will tolerate Zone 5, but most do best in Zones 6 to 8. They also dislike a lot of wind. They should be planted in partial shade or in a sheltered spot in sun.

Japanese maples are especially valued for their decorative foliage, which may be green, golden, or purple, and may be finely divided or typically maple-leaf shaped. Most Japanese maples grow slowly to a height and width of about 20 feet, but some, such as the species *dissectum,* mature to only about half that size. Of course, all are suitable in a garden that has a Japanese style, but they can improve the appearance of almost any garden corner or spot by a patio or path. Their natural elegance should not be compromised by coarser plants nearby. They are natural companions for large rocks, mosses, and grasses. The smallest are comfortable in a large pot. Most offer fall color.

Although Japanese maples are often raised in dry western gardens, summer heat causes the leaf tips to die, giving the trees a tan cast and muting the splendid fall color.

TREES, CONIFER

Large Fir

White fir
Abies concolor

Firs are trees for large estates or country properties. They grow so tall and narrow, to about 50 feet, that they are unsuitable for small gardens, where they will soon look out of proportion. They do best given plenty of space in a fairly cool climate. In the right spot—at the back of a sizable property, behind smaller deciduous trees, or along a boulevard—a mature fir can be dramatic. The soft, bluish green needles are attractive all year.

The white, or Colorado, fir is more tolerant of a variety of growing conditions than most other firs. It will put up with sun or light shade, dry soil or moist. It can be grown in warm or cold areas, to Zone 4, preferably in the Midwest, North, and East. Unlike other firs, it will tolerate city conditions. Its only dislikes are prolonged drought and excessive heat.

Large Pine

White pine
Pinus strobus

The white pine is a beautiful blue-green native American tree that you should choose only if you have a large estate or farm. Even there, this tree can eventually dominate the landscape. It is long-lived and majestic, growing 50 to 75 feet tall in 40 years and perhaps not stopping until it reaches 150 feet. If you want a smaller tree, consider one of the cultivars, such as 'Contorta', which has attractively twisted branches and grows less than 20 feet tall.

White pine is best planted on its own, as a specimen, so its handsome shape can be appreciated—pyramidal when young and plumed when older. If you have a large property, white pine also makes a good forest tree. Pines are generally more accepting of climatic differences than spruces or firs. Though white pine does best in well-drained soil, it will survive in a wide variety of soil types, from dry to boggy in partial shade or, better, full sun. White pine can survive as far north as Zone 3.

Left: White fir, one of the stateliest of evergreens, is not a tree for the small yard. It is most attractive when seen from a distance across a broad lawn. Right: White pine is an American native that takes well to garden conditions.

Left: Many varieties of blue spruce are available, with colors that vary from dark green to an almost silver-blue, and in forms that range from towering to sprawling and limp. Right: Scotch pine is a tough landscape tree also valued as a Christmas tree because of its regular pyramidal shape when young.

Large Spruce

Colorado blue spruce
***Picea pungens* 'Glauca'**

Because of its distinctively colored foliage and absence of problems, Colorado blue spruce is extremely popular in gardens. If you want an evergreen with bluish foliage, there is no better choice for a large yard.

Think twice, however, before buying a Colorado blue spruce as a specimen for the middle of the lawn or next to the house. Not only will your garden probably look like every other garden on the block, but also your tree will not be shown to its full advantage. Colorado blue spruce shows off its color best when it grows in close company with other plants. It should be contrasted with dark green conifers or grayish, reddish, or green deciduous trees behind or beside it, and with green shrubs in front.

Colorado blue spruce grows best in Zones 2 to 7. It forms a neat cone shape, broad at the base and as tall as 100 feet, though it is slow growing and can be kept smaller with pruning. It is tolerant of most soils and growing conditions but does best in moist soil in full sun.

Medium Pine

Scotch pine
Pinus sylvestris

Many people know Scotch pine best as a blue-green Christmas tree. When young, it has a typical triangular shape with evenly spaced, horizontal branches, but as it ages the lower branches die, leaving an attractive spreading, drooping umbrella of upper branches. The bark on older trees has a beautiful orange tint. Trees can grow as tall as 75 feet, but they can be kept much shorter by pinching back the candles of new growth by one-half in spring.

Scotch pine is one of the best evergreens for poor, dry soil. It will tolerate any soil that is well drained. It will also put up with strong winds. It is hardy from Zones 2 to 8, though it grows better in cooler areas.

Medium Yew for the North

Japanese yew
***Taxus cuspidata* 'Capitata'**

Yews have flat, often glossy, soft, evergreen leaves that add a note of graceful calm to almost any garden. The Japanese yew is much easier to take care of than its relative the English, or common, yew, *T. baccata.*

Most garden yews are shrubs, but 'Capitata' has a strong central stem with attractive, grayish brown bark. Japanese yews can reach 45 feet, but they grow slowly and will withstand shearing or pruning to keep them smaller and more tidy.

Japanese yew is problem free provided it is planted in soil with excellent drainage. It grows in Zones 4 to 7, but in the northernmost reaches of its range it needs a protected place, perhaps close to a wall. Cold, dry winters and strong winds can cause the leaves to turn brown. In the southernmost extension of its range or where summers are hot and dry, it should be given some shade.

Medium Yew for the South

Yew pine
Podocarpus macrophyllus

Not a true yew but looking like one, the yew pine can provide welcome medium green color for Southwest and Gulf Coast gardens in Zone 8 and warmer. It looks especially attractive combined with ferns and shade-loving tropical plants. This Chinese species grows 10 to 30 feet tall and about half as wide, forming a tall oval or column. It is easily pruned or trained.

Yew pine grows easily in any soil, provided it is not constantly wet, and accepts light shade to full sun. It should be sheltered from strong winds. The foliage can be burned by sun, wind, or salt spray when it is young, but it toughens with age.

Yew pines are good candidates for large tubs. They are best staked after planting and until growth resumes.

Top: Yew pines, neither pines nor yews, look like large-scale yews. They can be shaped into hedges and topiary like yews, but are most attractive when allowed to develop their own natural form. Bottom: Dwarf Alberta spruce is the basic foundation plant of many a garden. Naturally tidy, it is amenable to being shaped into balls, columns, or hedges. It is particularly attractive when displayed, as here, with rocks.

Small Conifer for the North

Dwarf Alberta spruce
***Picea glauca* 'Conica'**

This variety of white spruce may be sold as dwarf Alberta spruce, Alberta spruce, or dwarf white spruce. It is a natural dwarf that grows very slowly, only about 2 inches a year, to an eventual height of 10 feet at most, though it can be kept shorter by shearing. The dense foliage is bright green, and the tree maintains a compact, formal, pyramid shape whether or not it is pruned. It is hardy to Zone 1.

This small tree can be used to punctuate the ends of a foundation planting, to edge a

The sprays of Hinoki false-cypress give it an interesting sculptured texture reminiscent of growths of coral or lichen.

path or driveway, or to provide year-round color in northern perennial borders.

It does best if planted in moist loam in full sun, but it tolerates poor, dry soil, wind, and some shade.

Small Conifer for the South

Hinoki false-cypress
***Chamaecyparis obtusa* 'Nana'**

Where a small, coniferous evergreen is desired, this cultivar of Hinoki false-cypress is ideal almost everywhere in Zone 7 or farther south. It forms a neat pyramid that grows slowly to about 6 feet tall, making it useful for city yards, rock gardens, perennial borders, foundation plantings, and large containers. After 10 years, it may be only 2 feet tall. The dense crop of leaves, formed in shell-shaped sprays, are dark green above, whitish beneath.

False-cypress does best in full sun and rich, somewhat moist, well-drained soil. Given the conditions it prefers, it should be problem free.

SHRUBS, BROADLEAF

Large Shrub With Fragrant Flowers for the North

Lilac
***Syringa × hyacinthiflora* 'Excel'**

Lilacs are loved for their wonderfully fragrant spring flowers. Later in the season they are quite unassuming in shape and foliage color, however, so they are best placed behind other shrubs or perennials that pick up the limelight after the lilac flowers fade.

The hybrid 'Excel', which grows 10 to 12 feet tall, is hardier and flowers earlier than its parent, the common lilac, *S. vulgaris.* Also to its advantage, 'Excel' is more disease resistant than other lilacs and does not produce suckers. It can be grown from Zone 2 to about Zone 6; like most lilacs, it needs cold winters to bloom fully. It has pink flowers, which bloom from top to base. For best flowering, it should be planted in sun, and the flower stems should be pruned off after blooming to encourage bushiness and next year's bloom. For purple flowers in the North, consider the cultivar 'Assessippi'. In more southerly gardens, choose the Descanso Hybrids, cultivars of *S. × laciniata.*

Large Evergreen for the North

Holly
***Ilex × meserveae* 'Blue Prince', 'Blue Princess'**

Mrs. F. Leighton Meserve, a New York gardener, made evergreen hollies possible in gardens as far north as Zone 5 by crossing two *Ilex* species. Like all hollies, Meserve's creations produce berries only on the female plants, and both a male and female plant are necessary for pollination. 'Blue Prince' is, of course, the male of this particular couple; 'Blue Princess' is the female, which produces a full crop of red fruit if there is a 'Blue Prince' within about 20 feet. Both have lustrous, dark green foliage. 'Blue Prince' grows about 12 feet tall, his mate a few feet taller. One male, which will serve several females, is all that is necessary if you want to grow a hedge or a group.

These hollies should be planted in partial shade in a protected place where they will not

Top left: 'Excel' lilac displays trusses of lavender-pink flowers in the spring.
Top right: Many people grow holly for the bright winter berries. 'Blue Princess', a female holly, bears only if a male holly is present, but it is not necessary to plant a male if one grows in your neighborhood: pollinating bees are no respecters of property lines.
Bottom: Tea olive blooms in fall and winter in the South, where it scents gardens at a time when fragrance is rare.

be exposed to strong winds, especially in winter. They prefer ordinary or slightly acid, well-drained soil but do not tolerate long periods of drought.

Large Evergreen Shrub for the South

Tea olive
Osmanthus fragrans

For warmer gardens where the temperature never dips below zero—from Zone 7 south—the tea olive is both easy and beautiful. Also called sweet olive because of the Chinese custom of using the leaves to flavor tea, this shrub will withstand drought and many kinds of adverse conditions. The tea olive blooms in spring in Zone 7, but farther south it blooms in winter and continues for several months. The small white flowers have an intoxicating perfume, suggestive of apricots.

Tea olive grows as tall as 20 to 30 feet and may assume the shape of a small tree where it

Top: The compact form of winged euonymus is called "Burning Bush" for its brilliant scarlet fall color, brighter than any other plant. Bottom: Grown for its red berries as well as its graceful foliage, heavenly bamboo is also available in a dense dwarf form.

is content, preferably in neutral or slightly acid soil and partial shade. Grown in a pot, it will remain smaller. It makes an excellent potted plant that can be brought indoors for the winter. Kept moist and somewhat cool near a bright window, it should bloom indoors.

Large Shrub With Red Foliage

Purpleleaf sand cherry
Prunus × cistena

A cross between the cherry plum and a wild cherry, purpleleaf sand cherry has beautiful, reddish purple foliage, which makes it especially valuable in northern or mountain gardens where few other red-leafed shrubs flourish. Hardy and problem free, it makes an excellent accent shrub beside a path or stairs or as part of a foundation planting. The fragrant, pale pink blossoms come in spring and are followed by small, dark purple fruit.

Sand cherry does best in well-drained soil and full sun. It can reach 7 to 10 feet tall and almost as wide, and may become ragged looking, but pruning once a year will encourage it to remain smaller and neater. It is hardy from Zone 3 south.

Large Shrub With Winter Interest

Winged euonymus
Euonymus alatus

Consider the winged euonymus for shade or sun, as a specimen, or in the company of other shrubs. This popular plant, named for unusual corky wings on the branches, grows rather slowly into a stiff, rounded shrub 8 to 10 feet tall and wide, though it can be kept smaller. In fall, there is a crop of pretty but inconspicuous red fruit that resembles bittersweet. More obvious is a show of rose-colored to brilliant red foliage. The wings and the brown and green bark make it a valuable plant in winter, when it continues to look handsome.

Winged euonymus is problem free and widely adapted from Zones 3 to 8 in sun or shade, though it does not do well in very dry or waterlogged soils.

Medium Bamboo

Heavenly bamboo
Nandina domestica

Also called sacred bamboo, this is not a bamboo at all, though its airy, vertical stems and slender leaves suggest that plant. Heavenly bamboo is easy and effective in gardens as far north as Zone 7. Its graceful, compound leaves remain evergreen in the South; farther north they turn a beautiful red in fall. The bright red berries that remain on the branches in winter are the main attraction of this plant. They ripen in fall and are not hidden by the foliage, so they are very showy.

Heavenly bamboo is effective grown in groups, as a specimen surrounded by lawn, or behind broadleaf or coniferous shrubs. It tolerates either sun or shade but needs a spot sheltered from strong winds. It grows quickly to 6 to 8 feet tall. Plants that have become leggy should be cut close to the ground during winter; fresh, new shoots will arise in spring. If you want shorter stems, select 'Harbour Dwarf', which grows 2 to 3 feet tall and is well suited to confinement in a pot.

Heavenly bamboo loses its leaves at 10° F and is killed to the ground at 5° F, but in protected places north of Zone 7, it may surprise you by reappearing in spring.

Medium Shrub With Pink Flowers for the North

Smoke tree
Cotinus coggygria

Very distinctive while in bloom, the smoke tree, or smoke bush, is named for its flowers, which are so tiny and hairy and numerous that the shrub looks like a hazy cloud from July through September. It can be used as a specimen plant surrounded by lawn, but it looks better at the back of a perennial border or behind other, lower shrubs that provide interest when it is not in bloom. Green-leafed and purple-leafed forms, such as 'Daydream' and 'Royal Purple', respectively, also make good neighbors for an easy-care collection of shrubs.

Smoke tree is hardy to Zone 4 and tolerates any garden soil, even very dry or rocky, in sun or partial shade, though it does best in full sun. It grows 15 to 20 feet tall and equally wide if allowed to keep its multiple stems, so it should be given plenty of space. Its fall color is best where summers are dry and winters are cold. Depending upon the weather and the cultivar, the oval leaves turn yellow, red, purple, or orange in fall.

Top: Beautiful in bloom, smoke trees are also attractive after the blossoms fall: they keep their smoky color and texture, as hairs on the flower stems turn from pinkish gray to soft gray or purple.
Bottom: Glossy abelia is a hard-working shrub that remains in bloom all summer and into the fall. It can be left a rounded shrub or carved into a hedge or other shapes.

Medium Shrub With Pink Flowers for the South

Glossy abelia
Abelia × grandiflora

The most popular member of its genus, glossy abelia has shiny foliage, evergreen in the southern part of its range, and long-lasting masses of white or pinkish flowers, which bloom from early summer until frost. It grows from Zone 6 southward, though it may not survive a particularly cold winter in the northern part of its range, and even after an average winter it may need some deadwood pruned out. It grows 6 to 8 feet tall and almost as wide, forming a rounded bush.

This shrub does best in well-drained, acid soil in full sun or partial shade. It should be watered weekly in dry summers. Plant glossy abelia in groups or as a hedge or a specimen plant, surrounded by lawn or bedding plants such as impatiens.

Top left: Japanese snowball's red berries keep it attractive all summer long.
Top right: The butter-yellow blossoms of Siberian peashrub complement the fresh green of the foliage.
Bottom: Warminster broom thrives in hot, dry regions with mild winters.

Medium Shrub With White Flowers

Japanese snowball
***Viburnum plicatum tomentosum* 'Shasta'**

Suitable as a specimen plant or within a group of smaller shrubs, Japanese snowball is an elegant flowering shrub for gardens in Zones 6 to 8. The common name describes its delightful, ball-shaped clusters of white flowers that bloom in May or June. On 'Shasta', these flower balls can reach 6 inches in width. This cultivar is unusually horizontal. It grows about 6 feet tall and twice as wide, so it is excellent shading for a patio or pond. Its flower balls are followed in summer by fruits that are bright red, gradually turning black. The toothed, dark green leaves turn reddish in fall. The shrub remains attractive in winter because of its horizontal habit and gray bark.

Japanese snowball does best in moist, well-drained soil in full sun or partial shade.

Medium Shrub With Yellow Flowers for the North

Siberian peashrub
Caragana arborescens

For dry, windy places with cold winters, such as the prairies, the Siberian peashrub is dependable enough to be used for hedges, windbreaks, and specimen plantings. This is a shrub to situate on the windward side of slightly more delicate plants. The pale green leaves are composed of tiny leaflets, in the manner of many other members of the pea family. The flowers are bright yellow and small but are borne in sufficient numbers to make a sunny show from May to September.

Hardy to Zone 2, this shrub tolerates any ordinary soil. It prefers full sun, though it will put up with some shade. It grows to about 20 feet tall but can be severely pruned if it becomes too large for its site.

Rose-of-Sharon 'Blue Bird' produces cool blue flowers in the middle of summer.

Medium Shrub With Yellow Flowers for the South

Warminster broom
Cytisus × praecox

One of the prettiest early spring-flowering brooms, Warminster broom is appreciated for its somewhat weeping habit, which results in cascades of pea-shaped, creamy yellow flowers in May. The shrub is bushy, with branches that arch outward when they approach their full length of about 5 feet, so the plant is as wide as it is tall. It is deciduous, but the dense cluster of green stems gives it an evergreen effect. It is hardy to Zone 7.

Broom can be planted in masses for best effect while in bloom, or it can be used on its own with other shrubs or behind perennials. It prefers sun. It does fine in poor soil, in part because it fixes its own nitrogen supply, but the soil should be well drained. To encourage flowering, it should be pruned back in summer.

Medium Shrub With Summer and Fall Bloom

Rose-of-Sharon, Shrub althea
Hibiscus syriacus

This popular, old-fashioned shrub is available with blue, white, pink, or lavender flowers. It blooms from summer into fall, when little else is in bloom. Because the foliage is not as attractive as that of other shrubs, you may want to put it in a container so it can be moved to the foreground when in bloom. Another option is to bank it behind smaller, spring-blooming shrubs that will mask it until it commands attention as its large, trumpet-shaped flowers begin opening. A light shearing each spring will keep it more dense and promote more flowers, but isn't necessary.

Rose-of-Sharon is hardy to Zone 5 and is tolerant of a wide range of conditions, including city pollution. Grow it in full sun or light shade. It will grow to 10 or 12 feet high, but can easily be kept lower with pruning.

Medium Shrub With Winter Interest

Bayberry
Myrica pensylvanica

Although bayberry does best in sandy or gritty soil, even in clay or salty soil it is an easy plant. It adapts to places that are cold or hot, dry or wet, sunny or partly shady. This species, which grows 3 to 8 feet tall, is native to the East Coast from Newfoundland to Florida—Zones 2 to 8. It is grown primarily for its foliage, which is dark green and hairy, with the distinctive fragrance of bayberry when crushed. The foliage lasts on the shrub until early winter, when the show is carried on by the fruit. The small, waxy gray berries persist all winter, contributing to

Top left: Bayberry displays gray berries. Top right: Goldflower produces bright yellow flowers all summer. Bottom: Woadwaxen is a summer-flowering shrub with yellow flowers for the North.

the plant's considerable decorative effect. The fruit is produced on female plants only, which are the usual plants sold commercially.

Bayberry combines beautifully with broadleaf or coniferous evergreen shrubs in a border. It is also effective in foundation plantings or behind perennial flowers. *M. cerifera,* a more southerly species damaged by temperatures below freezing, yields the berries used in making bayberry candles.

Small Evergreen Shrub

Goldflower
Hypericum* × *moserianum

For southerly places, Zone 6 or warmer, goldflower is a neat and dependable producer of large yellow flowers from midsummer until fall. It grows only 1 to 2 feet tall and about half as wide, so it is ideal for massing under trees or brightening a slope where other plants do not thrive. It can also be used as a specimen, surrounded by lawn. It should be cut back to the ground in winter.

Goldflower does well in dry or rocky soil in full sun, though it will put up with some shade and a variety of soils.

Small Shrub With Yellow Flowers

Woadwaxen
Genista tinctoria

Ideal for perennial borders and city gardens, even if the soil is poor and dry, woadwaxen grows just 2 to 3 feet tall and wide, forming a neat, rounded shape. The leaves are bright green; yellow flowers bloom in summer. Woadwaxen does best in hot sun in Zones 4 to 7.

CONIFEROUS SHRUBS

Medium Round Shrub

Mugo pine
Pinus mugo mugo

No taller than 8 feet and about twice as wide, with dark green foliage, mugo pine is valuable for foundation plantings, as an accent in a perennial border, or even in a container.

Mugo pine does best in sun or partial shade and in moist soil, although once established it will tolerate considerable drought. It grows slowly and can be kept shorter by pinching back the new green shoots about an inch every spring. It grows best from Zones 2 to 7.

Medium Vertical Shrub

Juniper
***Juniperus chinensis* 'Mountbatten'**

Less disease susceptible than some other junipers, 'Mountbatten' is a good choice at the corner of the house, by a front stairway, or as an accent plant at the end of a bed of shrubs. It forms a narrow pyramid, eventually about 12 feet tall. The foliage is gray-green and can be sheared once a year to keep it looking neat and to keep the shrub shorter if you wish.

Junipers do best in full sun; they tend to become ragged looking in shade. 'Mountbatten' prefers moist soil but will survive considerable drought; it is hardy as far north as Zone 4.

Top: Mugo pine remains a cushion all its life. It is especially attractive planted next to rocks.
Bottom: 'Hetz Midget' dwarf white cedar is smaller than mugo pine, but has the same globular shape.

Small Cedar

White cedar
***Thuja occidentalis* 'Hetz Midget'**

White cedar, also called Eastern arborvitae or American arborvitae, does best in deep soil but otherwise puts up with most conditions—sun or shade, moist soil or dry. It tolerates pruning and strong wind. Although it grows from Zones 3 to 8, it is less vigorous in the southern end of its range.

'Hetz Midget' naturally forms a dense globe shape without pruning, so it is excellent as an accent at each side of a stairway or on both sides of a path. It grows 3 to 4 feet tall and equally wide.

Small Juniper

Juniper
***Juniperus sabina* 'Broadmoor'**

All junipers are valuable landscape plants for perennial borders, rock gardens, foundation plantings, shrub plantings, containers, or near deciduous trees where the junipers can receive some sun. The cultivar 'Broadmoor' grows slowly in the form of an arch, maturing at about 2 to 3 feet high and 10 feet wide.

'Broadmoor' does well in sun and well-drained, even dry, soil. Although sun can cause winter browning, the plant will recover in spring. This cultivar is resistant to juniper blight and is hardy to Zone 4. It has a pleasant, characteristic fragrance.

Top left: Japanese privet is a useful hedge plant for the southern states.
Top right: This germander is unpruned to show its flowers. Pruned, it can be shaped into a very tidy small hedge.
Bottom: Meyer lilac is slower-growing and easier to control than other lilacs, but shows the same beautiful flowers.

DECIDUOUS HEDGES

Tall Hedge for the South

Japanese privet
Ligustrum japonicum

In gardens from Zones 7 to 10, Japanese privet is much easier to care for than the common, or English, privet, *L. vulgare.* Although less hardy than the common privet, in its range Japanese privet can even be planted along the north side of a building where few other plants grow well. It bears shiny, dark green leaves from the base to the top. If the flower buds are not sheared off, the plant produces a crop of creamy white flowers followed by black fruit. It grows 12 feet high and half as wide, although if left unpruned it may grow to 18 feet.

Japanese privet does well in sun or shade and in any soil except one that is always wet.

Medium Hedge for the North

Meyer lilac
***Syringa meyeri* 'Palibin'**

The common lilac, *S. vulgaris,* can be invasive—this is the plant that has colonized abandoned fields in the North—but the Meyer lilac is smaller and easier to control. Meyer lilac grows only 4 to 8 feet high and somewhat wider. Pruning should be done after flowering, so the flower buds are not cut off. Somewhat less hardy than the common lilac, the Meyer lilac can be grown in Zones 3 to 7.

This plant grows in any soil. For the best effect, it should be planted in front of evergreens. For the best flowering and disease resistance, it should be grown in full sun.

Medium Hedge for the South

Myrtle
Myrtus communis

In warm places, Zone 7 and southward, myrtle is a useful plant for hedges and shrub borders. It forms a rounded shape 6 to 10 feet high and almost as wide. The pointed, glossy, dark green leaves, evergreen in winter, are aromatic when crushed or brushed against. In summer, small, perfumed, white flowers bloom, followed by half-inch blue-black berries. Myrtle can be used as a formal or an informal hedge.

It does best in well-drained soil in sun. It is lovely along a walk, where the aromatic foliage can be appreciated.

Medium Red Hedge for the North

Purpleleaf sand cherry
Prunus × cistena

See the description on page 30. Purpleleaf sand cherry makes a dramatic hedge that can be kept 4 to 5 feet tall with annual pruning.

Low Evergreen Hedge

Korean boxwood
***Buxus microphylla koreana* 'Winter Green'**

A popular hedging plant, Korean boxwood produces dense green growth as far north as Zone 5. This cultivar is somewhat hardier than the species. It grows 1 to 2 feet tall and wide in the northern end of its range and twice as big in the South. It can be left as it is or sheared every spring for a more geometric appearance.

Korean boxwood does best with cool roots, so it should be given a mulch of lawn clippings or fallen leaves. It grows well in sun or shade in a place sheltered from strong winds, especially in winter. For the first month after planting, it should be protected from full sun, perhaps by inserting leafy twigs into the ground on the southern side of the hedge.

Very Low Evergreen Hedge

Germander
Teucrium chamaedrys

The classic low hedge of herb and knot gardens, germander is also excellent for defining a bed of roses or fronting a planting of medium-sized evergreen shrubs. It grows about a foot tall and twice as wide. The toothed leaves are dark green and aromatic when crushed. Rose or reddish purple flowers appear in summer. After flowering, the plant should be cut back.

Germander does well in sun or partial shade in any soil except one that is heavy and wet. It is hardy to 0° F, about Zone 6, although in the northern part of its range it may die back in winter and regrow in spring.

CONIFEROUS HEDGES

Tall Hedge for the North

Canadian hemlock
Tsuga canadensis

In all but hot, dry, windy gardens, Canadian hemlock can form a beautiful hedge so thick that nothing will grow underneath or pass through. A single plant has the potential of becoming a tall tree, but it grows slowly and can be kept 4 to 6 feet tall if pruned every year.

Canadian hemlock is hardy in Zones 4 to 7 but should be confined to places with acid soil. It needs conditions that are mostly cool and moist. It suffers in drought and strong winds, and it is in danger of being scorched if the temperature rises above 95° F.

Tall Hedge for the South

California incense cedar
Calocedrus decurrens

Named for the fragrance of the crushed foliage, California incense cedar is a naturally tall, slender tree, so you can create a hedge or screen simply by setting plants several feet apart in a straight line. This cedar can be dramatic along a driveway or at the back of a large property. Although the tree can grow as tall as

Although it grows to more than 60 feet tall in the forest, Canadian hemlock can be kept in a hedge with annual pruning.

70 feet, a point at which it might better be considered a windbreak or screen, it usually stops growing at 30 to 50 feet high and just 8 to 10 feet wide.

California incense cedar grows well in Zones 5 to 8 in sun or light shade. It withstands heat and drought. Although it prefers moist, well-drained, fertile soil, it will tolerate many kinds of soil.

Medium Cedar

Arborvitae
***Thuja occidentalis* 'Techny'**

Native to North America, this arborvitae, or white cedar, which is hardy to Zone 3, is the most frequently used hedge plant in the Northeast. This cultivar, also called 'Mission', has the advantage of remaining green in winter, unlike the species, which turns brown. An arborvitae hedge can be kept from 3 to 6 feet tall, but it should be trimmed by pruning off branches at their base rather than by shearing.

Although tolerant of all kinds of soil, it prefers moist, alkaline soil. It should be kept watered until it is established. Since it is damaged by strong winds, it is not suitable as a windbreak.

VINES

Large Vine for the North

Boston ivy
Parthenocissus tricuspidata

Tougher than English ivy in almost every respect, Boston ivy is a more refined relative of Virginia creeper (*P. quinquefolia*). It clings to anything, even stone or brick walls, and can grow 6 to 10 feet in a season, reaching a mature height of 40 to 60 feet. It forms a dense cover of somewhat ivylike leaves on walls and turns a beautiful red in fall. It is hardy to Zone 4.

Boston ivy puts up with any soil in sun or shade, and it tolerates wind and air pollution. This is the plant to cover exposed walls where nothing else will grow.

Large Vine for the South

English ivy
Hedera helix

This European species is used extensively as a ground cover and to climb a variety of surfaces. Most valued as a vine for shade, ivy will also tolerate sun if it is not too hot. It can swiftly

Left: 'Techny' arborvitae is one of the most popular hedging materials in the Northeast. Right: Climbing hydrangea is covered with softly delicate white flowers in spring, then turns pale gold in fall.

reach 80 to 90 feet given ideal conditions of rich soil, low wind, some shade, and temperate weather. It is hardy from Zone 6 south. In the northern part of its range, it does best if kept out of full sun, which can burn it in winter.

Ivy will grow in any soil. After planting, it should be kept watered until growth resumes. It may need considerable pruning back if it is content.

Large Flowering Vine for the South

Climbing hydrangea
Hydrangea petiolaris

Climbing hydrangea can grow 50 to 60 feet tall. The sweetly fragrant white flowers, which bloom in summer, are not as showy as the pale gold leaves in autumn. During winter, the reddish stems provide interest.

Hardy from Zone 5 south, it grows in any ordinary garden soil. In the northern part of its range, it should be situated on a north- or east-facing wall to avoid sun damage. It may grow slowly for the first year or two.

Medium Flowering Vine for the North

Silver lace vine
Polygonum aubertii

Not suited to a small garden because of its almost boundless energy, silver lace vine does best in a place where it can ramble. Attractive from spring to fall, it can grow 10 to 15 feet in a season; it dies back to the ground in winter. Clusters of flowers are produced in late summer.

Silver lace vine does well in any soil in full sun or shade. This is the vine to grow in difficult situations, even in Zone 4. In good soil, it can become weedy.

Annual Flowering Vine

Morning glory
***Ipomoea* spp.**

Morning glories are easy and dependable. Sow seed as soon as frosty nights are past; the vine should sprout within a week, reaching a height of about 8 to 10 feet by the end of summer.

There are cultivars that produce mixed colors, or you can choose just one color—blue, white, pink, purple, or red—or variegated. Flowers, as wide as 4 inches, open in the morning and last just a day; new buds are produced until frost. The vine will twine around a support or will clamber prettily over a wall.

Silver lace vine grows rapidly, producing foamy masses of white flowers. It blooms all season long.

Morning glories are an old favorite. Use them to hang over a wall—as these do—cover a stump, or twine up vertical strings outside a window to keep out the sun.

The Easiest Flowers

Beautiful but not fussy, dependable but not demanding, these flowers give the easy garden its brightness, color, and fragrance.

Flowers are the most dynamic plants you can grow. They may change from day to day, certainly from week to week. Whatever flowers you choose, they will be the plants that attract the most attention.

Annuals can usually be purchased as bedding plants in spring. However, a few are quite easy from seed if you have the patience.

Most biennials and perennials are at their showiest for less than a month. Interplant them or front them with annuals and bulbs, so there will always be something colorful in the garden and so perennials past their prime will be hidden.

Very popular as houseplants in the Victorian era, ferns are now gaining a new following as easy garden plants, especially for shade. They provide restful foils for plants with bolder leaves, such as hostas, or for more colorful annual or perennial flowers. They are also useful in flower arrangements and in foundation plantings. Almost all do best in dappled shade in damp, acid soil. Ferns require little care. They should be planted while dormant—from late fall to early spring. Make sure the roots don't dry out in transplanting.

The key to making a rose garden easier is in finding varieties that are disease resistant and hardy enough for your climate. Roses need fertile, well-drained soil and a spot that receives at least 6 hours of sun a day but is not too windy. In cold-winter areas, the graft—the bulge on the stem where the named variety was grafted onto the roots—must be at least 2 inches below the soil surface, and the plants should be mounded with soil, straw, or pine boughs. Almost all require at least a light annual pruning.

Cranesbill, a hardy geranium, is an eminently satisfactory perennial; it remains neat without pinching or pruning, has a long bloom season, and is free of pest and disease problems. See page 51.

CHOOSING FLOWERS

There are several types of flowers you can include in your easy garden. Annual flowers live just one growing season but make up for their short lives by blooming profusely, often for weeks at a time. Biennials produce a clump of foliage during their first year, flower the second year, and then die. Perennials may last many years. Their blooming season is often short, but many of them have attractive foliage to offer beyond their blooming time. Perennials have the advantage of not needing replacement every year.

When you look through the following lists, choose flowers for your climate zone by size, the color scheme you want, and then by the plants' requirements for soil, sun, shade, heat, and water. Choose plants suited to what your garden offers, rather than changing your garden to suit the plants. The following flowers are classified by color, but all of them can also be found in white, except as noted.

Like their vegetable-garden relative, annual sunflowers—this is 'Color Fashion Mixed'—are large, enthusiastic, and friendly. Fill back corners with their cheerful exuberance.

ANNUALS

Very Tall Yellow, Rose, or Red Annual

Sunflower
Helianthus annuus

The most famous cultivar of sunflower is 'Russian Giant' or 'Giant Mammoth', which may top 10 feet and bears huge flowers with edible, oily seeds loved by squirrels and birds. This top-heavy giant is best confined to the northern boundary of a vegetable garden. Other sunflower cultivars, some as short as marigolds, are prettier for home gardens and make excellent, long-lasting cut flowers. Almost all have dark brown centers and tolerate hot, dry conditions. 'Color Fashion Mixed' or the similar 'Sunburst Mixed' grows about 5 feet tall and has a branching habit. Flowers are about 5 inches wide and are borne from midsummer until frost, with later flowers somewhat smaller than the first ones. Mixtures include all colors from the standard yellow petals to peach, bronze, orange, pink, and bicolors. Seed mixtures do not include white, but the cultivar 'Italian White', about 4 feet tall, is lovely.

Sunflowers are easily grown from seed sown in a sunny place as soon as the soil can be worked in spring. As the soil warms, seedlings sprout and grow quickly. They should be thinned to about 2 feet apart. Thinned seedlings transplant easily. As the flowers grow, staking may be necessary in windy places, but otherwise sunflowers have strong stems that are self-supporting. Sunflowers will self-sow if some of the flowers are allowed to dry on the stems.

Tall Pink to Scarlet Annual

Cosmos
Cosmos bipinnatus

Although cosmos is so easy that some gardeners dismiss it, the 3-inch, daisylike flowers are graceful and can be as lovely as large, single roses. The central stalk of cosmos, which is woody, grows about 4 feet tall. Staking may be needed in windy places. The long, slender flower stems are good for cutting. There are many cultivars, some bicolored and some with oddly shaped petals. 'Sensation', a mixture of

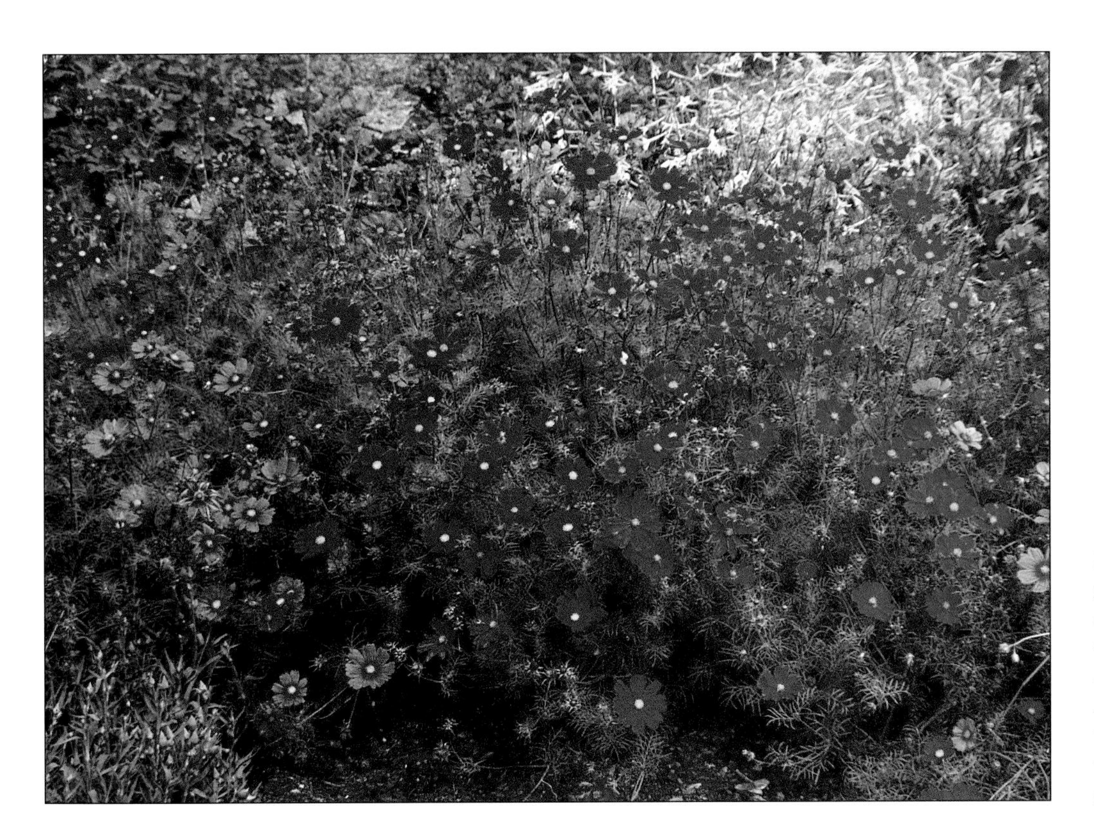

Top: Planted in a large group, 'Sensation' cosmos will attract attention from across the yard.
Bottom: In spite of its height, spiderflower is a delicate and gentle flower.

white, pink, rose, and magenta, is an older, dependable variety that will self-sow if some flowers are allowed to go to seed. Next spring, seedlings are easily weeded out or transplanted if you wish. There are also cultivars with a lower growing habit, such as the 2-foot 'Sonata'.

Cosmos is easily grown from seed sown in a sunny place as soon as the soil can be worked in spring. Seeds sprout as the soil warms, and seedlings grow quickly. You can also buy transplants in spring. Plants bloom from midsummer until they are killed by the first frost.

Tall Pink or Purple Annual

Spiderflower
Cleome hassleriana

The only garden member of the caper family, spiderflower looks different from anything else you are likely to grow. The flower is large, loose, and quite showy, with paddle-shaped petals and long, hairy stamens. Narrow, curving seedpods suggest spider legs. Woody stems about 4 feet tall make this a flower that looks best in groups at the back of a border. It is unusual enough to attract attention if grown close to a house wall or near a porch or door, where it can be seen easily. In a windy place, it may need staking.

Growing so low and dense that it can be used as a ground cover in a perennial bed or under roses, sweet alyssum makes a perfect foil for other flowers. In white, sweet alyssum brightens the colors of flowers planted near it.

Wax begonias are one of the few bedding plants that flower profusely in the shade. Along with impatiens, they provide color under trees better than any other flower.

Small White or Purple Annual

Sweet alyssum
***Lobularia maritima* cultivars**

An otherwise modest plant, sweet alyssum when massed produces a ground cover of almost uninterrupted color that is perfect at the front of a border or around taller flowers in a container. The flowers are lightly scented. The branches are about 10 inches long but are so soft and trailing that they grow only a few inches tall. Because of the length of the stems, sweet alyssum should be planted no closer than a foot from the front of a border.

Purple and white cultivars are available almost everywhere as bedding plants in spring. Pink, yellow, and peach cultivars can be a bit harder to find. Sweet alyssum is also easily grown from seed sprinkled over the soil and lightly raked in anytime from early spring until early summer. Flowers begin to bloom about 6 weeks after sprouting. Whites such as 'Carpet of Snow' are the most dependable and in most situations the brightest and showiest. White sweet alyssum complements any color of flower or leaf growing near it.

This is a plant for sun or shade, in almost any area of the garden. It will continue to bloom past the first fall frosts. Sweet alyssum often self-sows if you let it go to seed, but you may choose instead to shear off the seedpods to encourage more flowers.

Small Pink to Red Annual

Wax begonia
***Begonia semperflorens-cultorum* hybrids**

The wax begonia is a tender perennial commonly grown as an annual, although if brought indoors in a container in fall and sheared back, it will reveal its perennial nature with a winter show of flowers. It can then be sheared back and moved outdoors again in spring, or cuttings can be taken to propagate the plant.

Wax begonia began its cultivated history as a flowering houseplant. As such, it is a natural for shade, though it will bloom in full sun as long as the weather is not too hot. The flowers, in shades of red through dark to light pink to white or bicolored, are generally a little bigger than an inch wide. The attractive leaves, which look waxy, may be green, bronze, or variegated. Standard plants form roundish mounds 6 to 8 inches high. Hanging basket types have reclining stems more than a foot long. Many cultivars are available as bedding plants in spring.

Small Pink, Orange, or Purple Annual

Impatiens
Impatiens walleriana

The most popular bedding plant in America, impatiens comes in a range of pretty colors, never needs its spent flowers removed, blooms reliably from late spring until frost in full or partial shade, and always looks neat. For the best appearance, the soil should be kept watered during dry weather.

Impatiens can be purchased as bedding plants in spring. After the last spring frost, the plants can be set out about 8 inches apart, close enough so they are almost touching. The bare soil between them will be covered as they grow. Impatiens looks fine at the front of a border, massed as a ground cover under trees, or grown in containers in the shade. A grouped planting of a single color looks best. In fall, cuttings of favorite cultivars can be rooted in water and kept as houseplants until the following spring, in the manner of geraniums, described on page 45.

New Guinea impatiens (*I. hawkeri; I. petersiana*) grows about a foot tall and is more sun tolerant than the species, with dark green or variegated leaves.

Small Yellow, Pink, or Violet Annual

Moss rose
***Portulaca grandiflora* Cloudbeater hybrids**

Like little neon lights, moss rose, or portulaca, flashes some of the brightest colors in the garden. This recumbent plant enjoys hot, dry weather and is excellent anywhere you are not likely to water—beside a driveway or between the stones in a pathway. It is attractive planted at the front of a border in sun or in a shallow container for a sunny balcony. It is one of the few flowers that look just as good grown in a mixture of colors as in a single tint.

Flowers open in the morning and bloom until afternoon. In older cultivars, the flowers closed when the sun did not shine, but flowers of Cloudbeater hybrids stay open whatever the weather. Its flowers are about 2½ inches wide. Plants spread about a foot wide and grow 6 to 8 inches tall.

Moss rose can be purchased as bedding plants in spring, or seed can be sprinkled in the garden after the last frost. Flowers bloom about 8 weeks after sowing and may self-sow.

Moss rose looks best when planted, as this mix of Cloudbeater hybrids is, in large masses. It tolerates the hottest and driest of conditions. As with other succulents, the pigment that gives color to the blossoms is composed of a different group of chemicals than pigments in most flowers—it has a neon quality.

French marigolds are smaller and more compact than the African species. Use them for borders around flower beds or in small containers. This luscious orange is 'Safari Tangerine'.

The plants look best grown in masses, close enough so they will overlap. Their only demand is well-drained soil. This is a desert plant that does poorly with too much rain or prolonged dampness.

Small Orange, Yellow, or Bronze Annual

French marigold
Tagetes patula

The smaller marigolds include the so-called French marigolds, *T. patula,* which are often bicolored and may be delightfully single, such as the orange and mahogany 'Granada'. Another group of single marigolds with innocent charm is the Gem series of signet marigolds, *T. tenuifolia* (formerly *T. signata*), which are perfect in pots and window boxes. Height varies from 8 to 12 inches, and flowers are an inch or two wide.

French marigolds do well in sun or partial shade. They look best in masses of the same color, planted closely enough that they touch when full grown. The attractive, finely divided foliage is bushy to the ground, so these marigolds can be grown at the front of a border.

EARLY BIENNIALS AND PERENNIALS

Tall Multicolored Biennial

Foxglove
Digitalis purpurea

Biennials are flowers that take two years to bloom. That leisurely schedule makes them most suitable in woodlands or relaxed cottage gardens along with other biennials such as Canterbury bells (*Campanula medium*), hollyhock (*Alcea rosea*), and sweet William (*Dianthus barbatus*).

Foxgloves are usually, but not always, biennial. The tallest types, sometimes 8 feet tall, are most likely to spend their first year as a ground-hugging rosette of downy leaves, but shorter varieties such as the Foxy hybrids may bloom the first year. If the plants you buy have flower stalks, you'll know they are on their way to blooming that year.

Foxgloves appreciate moist, loamy soil in partial shade, but they will bloom almost anywhere provided they have shelter from wind and high heat. They should be grown at the back of a flower border or next to a wall where their height will not block shorter plants. Often they will self-sow and naturalize. Flowers are purple, red, pink, yellow, or cream.

Medium Blue Perennial

Siberian iris
Iris sibirica

The Siberian iris is less well known than its cousin the Dutch iris, yet this hardier species, to Zone 3, has many virtues. Chief among them is season-long, attractive foliage, about 3 feet tall, resembling a clump of wide-bladed, dark green grass. But the flowers are also beautiful—slender and more elegant than the Dutch iris, in clear shades of blue to purple.

Siberian iris does well in sun or partial shade, preferably where the soil is not too dry. They are among the few plants that enjoy wet soil, and although they will survive drought, they will not grow as large. They are best in the middle of the perennial border, where the flowers can be appreciated in late spring. When flowering is finished, the clump of foliage will

cover the stems of taller plants behind. Siberian iris is also excellent among shrubs in a foundation planting or near a pond.

Medium Yellow Perennial

Leopard's-bane
Doronicum orientale

The surprise with leopard's-bane is daisylike yellow flowers early in the season, when you least expect this dash of summer color. Leopard's-bane is an easy plant, especially where summers are cool. It is hardy as far north as Zone 4. It is tolerant of poor soil and will grow in full sun or partial shade.

Plants grow 1 to 3 feet tall and become dormant in summer, when there will be a blank spot of soil over their roots. They should be grown behind low, leafy plants that will camouflage their quiet season. Leopard's-bane looks best grown in groups of at least 3 plants set a foot apart.

Top left: Native to the Pacific Northwest, where it blooms under the cover of firs and hemlocks, foxglove tolerates more shade than most tall flowers, but it will also grow in the sun if the summer isn't too hot.
Top right: The hot-yellow flowers of leopard's-bane bloom in early spring.
Bottom: Siberian iris is a plant of boggy places; it will thrive next to a pond or in a wet spot. Its foliage is neat all summer long.

Medium Pink or Purple Perennial

Peony
***Paeonia* species**

Considering their exotic appearance, peonies are surprisingly hardy plants, surviving in abandoned gardens as far north as Zone 4. The flowers, which may be single or double, in shades from light pink to dark purple, are as wide as 6 inches. They bloom in late spring on shrubby plants 2 to 3 feet tall. Although flowering lasts only about 2 weeks, the dark green foliage persists and is attractive until fall, when it turns brown and should be cut back to the ground.

Peonies, with their lush, generous flowers, are old-fashioned plants reminiscent of Grandma's garden. This beautiful yellow is 'High Noon'.

Double peony flowers are so heavy that their stems often bend to the ground, especially after a rain. To prop them up, surround the entire plant in spring with a low tomato cage, a circular cage designed specifically for floppy plants of this height, or a circle of stakes and strings that will be hidden in the foliage as the plants grow. No additional care is required. Single peonies, whose flowers resemble large poppies, are better able to support themselves. Peonies look attractive grown along a driveway, as specimen plants, or in the middle of the perennial bed.

Small Perennial for Shade

Violet
Viola odorata

At the front of a border, violets are eyecatchers in late spring, when their blue, yellow, white, or purple flowers begin to bloom. Flowering can continue for weeks if the weather stays cool. The attractive leaves are roundish and glossy green. Violets are apt to self-sow if they are content, and can appear where you least expect them, but these are such classy flowers that they can be forgiven their spreading nature. They are useful to help fill spaces between other, taller things, from spring bulbs to summer annuals. Unwanted seedlings can be pulled out easily.

Violets should be planted in groups, with plants 4 inches apart, anyplace where their small size (about 6 inches tall) will be an asset.

MIDSEASON PERENNIALS

Tall Purple Perennial

Purple coneflower
Echinacea purpurea

Although purple coneflower has been celebrated recently for the medicinal, antibiotic value of its roots and flowers, gardeners have long appreciated this daisy's aesthetic value in the landscape. Native to North America, this attractive and distinctive plant has downward-slanting mauve petals and protruding orange-brown centers. The configuration of the flower has caused some confusion with the genus *Rudbeckia.*

Stems may grow as tall as 4 feet but are generally self-supporting. Purple coneflower does well in full sun or partial shade, in front of a wall, in the company of other tall daisies, or against a backdrop of evergreens that will complement its subtle colors. It does best with weekly waterings throughout the summer, although it will survive drought. It self-sows modestly where it is content. It can be grown in places as cool as Zone 3.

Medium Yellowish Perennial

Daylily
***Hemerocallis fulva* hybrids**

Daylilies have naturalized along country roadsides in the North, a sure sign of a perennial that can survive without human interference. In the garden, too, daylilies take care of themselves, yet they are beautiful all season, even in Zone 3. The flowers, in orange-toned shades from cream and yellow to pink, bright orange, and bronze, bloom just a day apiece, but there are so many that flowering continues for weeks, sometimes months. Even after the flowers are finished, the wide, grassy, dark green foliage decorates the garden.

There are hundreds of daylily cultivars, mostly with stems 2 to 3 feet tall. The tetraploids have large flowers so they are apt to need staking; avoid these if you don't want to take the trouble. Among the recently developed long-blooming cultivars, the most famous is 'Stella d'Oro', with bright yellow flowers on stems 18 inches tall.

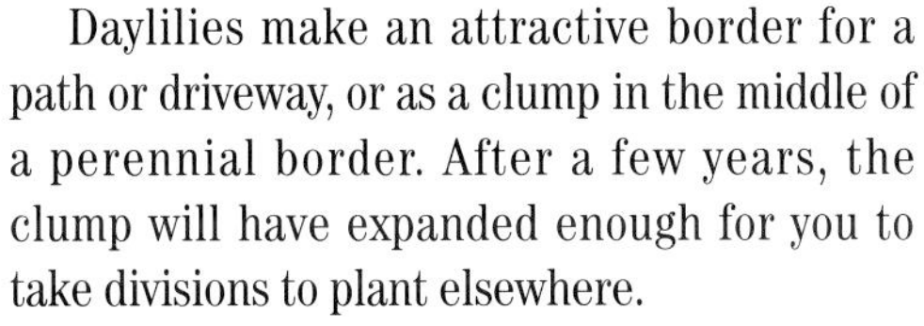

Daylilies make an attractive border for a path or driveway, or as a clump in the middle of a perennial border. After a few years, the clump will have expanded enough for you to take divisions to plant elsewhere.

Medium Purple Perennial

Salvia
***Salvia* × *sylvestris* 'East Friesland'**

Blue flowers suit almost every garden, but they are relatively rare. This dramatic hybrid contributes thin but densely flowered blue spires about 2 feet tall. The plant grows equally wide and is self-supporting. It prefers full sun and performs well even in dry soil. The grayish foliage is aromatic, reminiscent of sage. This salvia can be massed in the middle of a flower border or grown as a specimen on its own.

'East Friesland', also known as 'Ostfriesland', hardy to Zone 5, blooms in late spring to early summer, and sometimes again later in summer if the spent flower stalks are removed. It must be grown from plants or from division; it is not available as seed.

Top left: Purple coneflower is a plant of the tallgrass prairie of the upper Midwest. Extracts and powdered leaves, sold under its Latin name of Echinacea, *are said to strengthen the immune system.*
Top right: Perhaps the easiest of all perennials, daylilies will naturalize and thrive without human intervention. Although each flower blooms for only one day, new flowers open constantly. Many varieties bloom for months.
Bottom: Although related to the annual scarlet sage (Salvia splendens, *page 45*), *which has flaming red flowers, most perennial sages have blue or purple flowers. 'East Friesland', shown here, has an attractive clumped form.*

Small Perennial for Shade

Cranesbill
Geranium sanguineum

What most gardeners call geraniums are members of the genus *Pelargonium,* the floriferous South African tender perennials often grown in pots and as houseplants. The genus *Geranium* includes some very different plants. They are

'Little Miss Muffet' Shasta daisy blooms from June until frost. Plant it next to flowers with clear, bright colors; the bright white of the daisy will be like a spotlight on the other flowers.

hardy perennials, many native to North America, and their flowers are little saucers, followed by long, thin seedpods that give the plants the common name cranesbill. Many have blue flowers. On *G. sanguineum,* the flowers are bright magenta and bloom all summer. Plants form rounded mounds about a foot high, good at the front of a border or in containers in partial shade. Distinctive, divided foliage stays green until fall, then dies back to the ground. Cranesbill prefers well-drained soil.

Small White Perennial

Shasta daisy
Leucanthemum* × *superbum
'Little Miss Muffet'

Cultivated Shasta daisies are more restrained versions of the white daisies that bloom along country roadsides in summer. Some cultivars are doubles, with many more petals than the wild ones. Many grow so tall that they need staking, but 'Little Miss Muffet' (formerly *Chrysanthemum* × *superbum*) is a dense-growing, foot-tall cultivar. Shasta daisies do well in full sun or some shade. They grow in any ordinary soil, forming a gradually widening clump that can be divided in about 3 years.

LATE PERENNIALS

Tall Perennial for Sun

Russian sage
Perovskia atriplicifolia

Voted Perennial of the Year for 1995 by the Perennial Plant Association, Russian sage is adaptable to almost all conditions, will grow as far north as Zone 3, and is ornamental for a long season. It is very resistant to heat and drought and appreciates full sun; plants tend to be leggy and sprawling in shade.

This plant has an airy, shrubby form to about 4 feet tall. The foliage is grayish and sagelike. The blue flower spikes, which are pretty but not especially showy, appear in late summer and early fall and often continue blooming until late fall. Plants look best grouped on their own or toward the back of a perennial border.

Tall Perennial for Shade

Snakeroot
Cimicifuga foetida

Six-foot spires of white flowers grow from snakeroot, a North American native that is one of the few shade-loving perennials that grows tall. It looks best in groups rather than grown as single specimens. Plants should be set 2 feet apart near a shady wall or at the back of a shady perennial border. It needs moist soil.

Also called bugbane because of the odor of a related species, snakeroot is hardy to Zone 3. It flowers in summer in warmer zones, later in the cooler parts of its range. Blooming continues for about a month.

Medium Pink to Purple Perennial

Aster
***Aster* species**

Wildflowers native to North America, asters were sent to Britain for breeding and came back with a new name, Michaelmas daisy. Asters have small pink or light purple flowers on stems as tall as 5 feet, but the hybrids have shorter stems and larger flowers in a greater color range, from white to dark purple. There are all sizes, but the best are 2 to 3 feet tall. Examples are 'Alma Potschke', which has bright pink flowers, or the rose-colored 'Barr's Pink'. The taller ones need staking. Asters provide bright color late in the season when many other perennials are finished blooming and the annuals are fading.

Asters prefer well-drained but not dry soil and do well sun or partial shade. Some cultivars suffer from mildew, which causes darkening and falling of the foliage from the bottom up. Buy only disease-resistant cultivars. If you want your asters to continue blooming past about 3 years, you should dig up the clumps, discard the centers, and divide the remainder into several transplants.

EARLY BULBS

Tall Bulb

Tulip
***Tulipa* Darwin hybrids**

The Darwin hybrids are some of the largest tulips, about 2 feet tall, with the biggest flowers. Darwins are known as midseason tulips, indicating that they bloom in midspring. Some Darwins act like annuals rather than perennials, disappearing after a winter or perhaps two. Among the cultivars likely to last longer without replacement are 'Apeldoorn', 'Beauty of Apeldoorn', 'Golden Apeldoorn', 'Holland's Glorie', 'Oxford', and 'Striped Apeldoorn'.

All Darwins should be planted about 6 inches deep in a sunny place in fall before the first frost. They must have well-drained soil, not constant moisture. Where squirrels like to dig, cover the tulip bed with chicken wire camouflaged under a thin layer of soil. Tulips look best in groups of the same color, planted about 4 inches apart. Mixing them up results in a speckled effect.

After tulips finish blooming, cut the flower stems but leave the foliage in place until it turns brown and dries. The leaves manufacture food for next year's bloom. Tulips look best grown behind bushy annuals or leafy perennials, to screen the unattractive late-season foliage. If tulips are content, they will gradually produce a cluster of bulbs that can be separated and planted elsewhere.

Top: Russian sage is grown as much for its gray foliage and neat clumping form as for its late blue flowers. Bottom left: Aster 'Alma Potschke' blooms late in the summer, and into fall, bringing color in a season that suffers from a dearth of flowers. Bottom right: Darwin tulips are most attractive planted in clumps or masses. These are 'Dutch Fair' (yellow) and 'Viver' (orange).

Medium Yellow- or White-Flowering Bulb

Daffodil
***Narcissus* species and cultivars**

All daffodils and narcissus are members of the genus *Narcissus* and are therefore variations on the same theme. The former are best known for stems taller than a foot, bearing large, trumpetlike, brilliant yellow flowers, though there are also white and bicolored daffodils. Narcissus are smaller and often have white petals with cup-shaped centers that may be white, pinkish, yellow, or orange. The best are sweetly fragrant, such as 'Peridot' and 'Thalia'.

Daffodils should be planted 6 inches deep in fall well before the first frost; they need plenty of time to develop their roots. They need well-drained soil but will tolerate sun or light shade. They will naturalize if grown in grass that is not mowed until the daffodil foliage dies. Daffodils generally expand into a clump in several years, and look best that way. The clump can be dug in late summer and some of the bulbs can be pulled away to plant elsewhere.

Top: Daffodils naturalize in most parts of the United States, making them a completely care-free flower. Allow the foliage to die down naturally after blooming to fatten the bulb for the next year's blossoms.
Bottom: Grape hyacinth will bloom in light shade as well as sun, and is beautiful in clumps in a woodland garden as it pokes through the leaves in early spring.

Small Blue-Flowering Bulb

Grape hyacinth
Muscari botryoides

This small, 4- to 6-inch flower, named grape hyacinth for its resemblance to a bunch of tiny grapes, is charming massed along pathways, in containers, around trees, and between larger bulbs. The usual purple-blue color complements not only green foliage but also the colors of other flowers blooming nearby.

The corms should be planted 6 inches deep in sun or light shade. Unlike most other hardy bulbs, grape hyacinth produces a grasslike clump of foliage in summer, a surprise if you do not expect it.

Small Purple- or Yellow-Flowering Bulb

Crocus
***Crocus* giant Dutch hybrids**

Spring hardly seems complete without crocus blooms, suggestive of Easter baskets and the innocence of childhood. Crocuses are among the few hardy bulbs that look fine even when planted in a mixture of different colors, because all are pastels and whites that blend well. Consider crocuses under deciduous trees, in containers, along pathways, by the front steps, and along the edges of perennial borders.

Crocus corms should be planted pointed side up, about 4 inches deep in well-drained soil in sun. If they are happy, they will gradually form widening clumps. New colonies can be started by digging them up in late summer and planting some elsewhere.

MIDSEASON BULBS

Tall Blue-Flowering Bulb

Allium
Allium aflatunense

There are many alliums for the garden, flowering in a color range from yellow and white through pink and rose to purple. All have a vertical stem topped by an umbel of small flowers. One of the most dramatic and dependable is *A. aflatunense,* sometimes labeled 'Purple Sensation'. Its 3-foot stems produce 3-inch, dark purple umbels in early summer.

The bulbs are apt to multiply gradually where conditions suit them. Allium does best in reasonably fertile soil that can be dry or moist, in full sun or partial shade. It looks attractive behind and among tulips and irises to bloom about the same time, complementing those shorter flowers. Allium is hardy to Zone 4.

Tall Bulb

Lily
***Lilium* species and hybrids**

Although lilies are so showy that they look difficult to grow, given their few requirements they could not be easier. They need well-drained soil and a spot that is not too windy, but they will do well in full sun to light shade. Many are hardy to Zone 3.

Lilies vary in height from just over a foot to as large as 6 feet. Flowers are classed by the direction they face—up, down, or out. Plants can be purchased for spring planting, or the bulbs, which may be bigger than your fist, can be planted in fall. Lily bulbs never become totally dormant, so they should be planted immediately. They look best planted in groups of three toward the middle or back of a perennial bed, or in large containers. Lilies generally multiply if they are happy, and you can dig up the bulbs several years later in late summer to divide them. Also, new plants can be grown from the scales on the bulbs, or from the bulblets that grow up the stems. Break off a few scales from the outside of a bulb, and plant them in a sheltered spot.

This is 'Mona Lisa', an outward-facing oriental lily. Lilies tolerate more shade than most showy flowers.

Medium Bulb

Gladiolus
***Gladiolus* hybrids**

Easy and dependable, gladiolus can be a bit awkward in the garden, because its stems are so straight and rigid that it tends to grow at odd angles unless staked. Nevertheless it is a dependable source of bright color and big, exotic blooms, and makes a beautiful cut flower, much appreciated for indoor arrangements.

The large corms should be planted several inches deep in well-drained soil in a sunny place around the time of the last spring frost. The small white cormlets that develop around the corm can also be planted to grow to blooming size over a couple of seasons. Soon after the first fall frost, the corms should be dug, dusted off, dried outdoors for a day, then packed into paper bags and kept in a cool, dry place over the winter.

For flower arrangements, the stems can be cut when the lowest flowers are just beginning to open. The remaining foliage should be left in place to strengthen the corm for next year.

FOLIAGE ORNAMENTALS

Tall Ornamental With Broad Leaves

Castorbean
Ricinus communis

The castorbean grows rapidly from seed to about 6 feet tall in a few weeks of warm weather. With its huge, almost flat leaves and fat, rigid stem, this is a dramatic annual for the back of a flower border or for situating at each side of an entrance.

Around the time of the last spring frost, the big seeds should be planted an inch deep in a place sheltered from strong winds. Castorbean does well in full sun or partial shade. Because the seeds of castorbean are poisonous and are smooth and attractive, this is not a plant for gardens where there are children.

The large leaves of the castorbean give a tropical look to the garden. Castorbeans are poisonous. Because the smooth, shiny seeds look like candy, they are particularly dangerous to small children.

Medium Foliage Ornamental for Shade

Hosta
Hosta sieboldiana

This is the plant of choice for shady gardens. Hosta forms a cluster of leaves that may hug the ground or grow as tall as the 4-foot 'Frances Williams'. Colors vary from golden and cream to light or dark green, usually variegated; the leaves may be roundish or pointed, flat or pleated. Hosta does best in some shade in decent soil with regular watering. It often survives in dry soil, but growth will be slow.

Plants die back to the ground in winter, then emerge in late spring. They look best in masses—under shade trees, in front of walls as

Coleus is available in a wide variety of colors and combinations that will brighten a shady area as much as bright flowers.

foundation plantings, and along paths. Where they are content, they will gradually expand to form a solid ground cover. They can also contribute areas of green calm between brighter things in a perennial bed, or they can be mixed with coniferous shrubs for an easy-care border.

Small Foliage Ornamental for Shade

Coleus
Solenostemon scutellarioides

This tender perennial was known as a houseplant before it became popular for outdoors. Indoors, coleus established its reputation as a dependable plant for low light. Its flowers are insignificant, but the foliage can be as beautiful as any flower, in variegated patterns of green, pink, cream, yellow, and maroon.

Plants should be set outdoors after the last spring frost. They look best about 6 inches apart at the front of a flower border or along a path. They are also attractive grown in containers. In good soil and in a place where they do not dry out severely, they can grow about a foot tall and wide. In fall, before frost, favorite plants can be potted and brought indoors to winter over as houseplants.

Remaining neat and attractive all season long, hostas are a standby in shady perennial borders and beds. They can also be used as ground covers, and the dwarf varieties make good edging plants.

ORNAMENTAL GRASSES

Very Tall Ornamental Grass

Bamboo
Sinarundinaria nitida

The vertical, graceful attributes of all grasses are exemplified by bamboo. Some bamboos also share the shortcoming of many grasses—invasiveness—but not this Chinese native, sometimes called umbrellagrass. It can take as long as 7 years to reach its full height. It slowly forms a bluish or purplish clump, reaching as tall as 12 feet in places with cool summers but staying shorter where summers are hot. Umbrellagrass can survive winters as cold as –20° F, put up with full sun or a considerable amount of shade, and tolerate any soil that is not waterlogged or constantly dry.

Bamboo roots should not be allowed to dry out in planting; once planted, bamboo should be watered heavily for 10 days. After this, the plant should need little care except weekly watering during dry weather.

Tall Ornamental Grass

Eulaliagrass
***Miscanthus sinensis* 'Morning Light'**

Grasses can be surprisingly lovely in gardens. One attribute is their late flowering. They provide decorative plumes and seed heads in October and November when little else is attractive. The variety 'Morning Light' has magnificent, feathery silver plumes with a tinge of purple. It grows 4 to 8 feet tall, forming a gradually widening clump. The leaves have a white midrib and a white stripe along the edges. With the first hard frost, they turn their winter color of golden yellow or light brown.

This variety of Eulaliagrass prefers damp soil. It is excellent by a pool. The lower leaves can be torn off as they die back during the summer. This grass is hardy to Zone 5 and in protected spots to Zone 4. Where the growing season is less than 3 months long, eulaliagrass may not flower, but it is nevertheless worth growing for the slender, silvery foliage and decorative, dark stems. The old stems should be cut back in spring to make room for new growth.

Top left: Eulaliagrass makes a dramatic statement in the garden. Like many other grasses, it changes aspects with the seasons, and is even attractive in the winter.
Top right: As vibrant as fireworks, fountaingrass makes an arresting contrast to the placid serenity of a large stone.
Bottom: Valuable for its knobby appearance in a mass planting, as here, or as a tidy mound to contrast with exuberant flowers in a perennial border, 'Solling' fescue is useful in many landscape situations.

Medium Ornamental Grass for Sun

Fountaingrass
Pennisetum alopecuroides

Named for its clumps of leaves that arch outward, fountaingrass grows about 3 feet tall. It is topped in fall with long, russet-colored seed heads that look like feather dusters. The dead plumes can be left in place for winter interest. This is one of the hardiest species of pennisetum, capable of growing from Zones 6 to 9. In the warmer zones, self-seeding can be annoying, but this is not a problem in colder places.

Fountaingrass can tolerate a very hot, sunny position, and it will withstand drought, though strong winds can shred the leaves. Fountaingrass must be given plenty of room because of its arching habit, but it is well suited as a focal point in a perennial border or in a foundation planting of shrubs.

Medium Ornamental Grass for Shade

Tufted hairgrass
Deschampsia caespitosa

Well named, this grass produces a clump of foliage about 3 feet tall and wide, with a hairy inflorescence. It's worth growing for its foliage, especially in full or partial shade. It takes moist or dry soil. The foliage is rich dark green. This is a very hardy species, capable of living through –30° F winters without snow cover.

Small Ornamental Grass

Fescue
***Festuca cinerea* 'Solling'**

The fescues are among the hardiest of the popular ornamental grasses, to Zone 4. They suffered no winter injury in tests at the Minnesota Landscape Arboretum, where annual minimum temperatures may be as low as –30° F.

Related to the grass used for putting greens, the species *cinerea* forms a 6- to 12-inch-tall tuft of slender, evergreen bluish leaves, which complement other small perennials and bulbs in a border. It prefers sun and well-drained soil. All cultivars are good, but 'Solling' is recommended for its nearly nonflowering habit. Fescues look best in spring and summer, but 'Solling' looks better through the summer heat since it doesn't flower. Fescue flowers are not very showy.

FERNS

Tall Fern for the North

Ostrich fern
Matteuccia struthiopteris

The ostrich fern, named for leaves that resemble ostrich feathers, is perfectly adapted to the conditions of its native Northeast woods: rich, moist, acid, even boggy soil in partial shade. It does best where summers are cool; it can suffer in the South. It also looks best in spring, browning in late summer. It is attractive grown behind plants with dependable late-season color such as ornamental grasses. Ostrich fern assumes a narrow, vase shape that may grow as tall as 4 feet in optimal conditions, but in most gardens the fern will stop growing at about half that height.

Ostrich fern has a tendency to spread if content but can be contained by a mowed lawn or a pathway.

Tall Fern for the South

Royal fern
Osmunda regalis

This handsome species, named royal fern because of its large size, grows tall and narrow, so it should be planted in groups for best effect and kept to the back of a shady border or a spot in front of a wall. It can grow as tall as 10 feet, though it often stops at 3 to 4 feet.

Like most ferns, royal fern does best in fairly moist, acid soil in some shade. It dies to the ground in winter.

Medium Fern for the North

Sword fern
Polystichum munitum

Native to the American West, sword fern is hardy to Zone 4. It creates a dramatic clump, 3 feet tall and wide, of arching, glossy, dark green leaves. It is an evergreen—thus another name, Christmas fern—that does best in rich, moist soil, but it is quite adaptable and easy to grow. It looks best in the company of moisture-tolerant perennials.

Top: Ostrich fern fills in a shady corner. Bottom: Sword fern grows in the redwood forests of California, seeming most at home near trees.

Medium Fern for the South

Southern sword fern
Nephrolepis cordifolia

This evergreen fern is the one to grow from Zone 9 south. Southern sword fern will grow even in poor soil in light or heavy shade. It will expand into a clump 2 feet high and wide.

Small Fern

Polypody fern
Polypodium virginianum

This little evergreen fern, hardy to Zone 4, has more tolerance for dry soil than most other ferns. Because it can tolerate dry roots, this is a good choice for crevices between rocks or bricks. It grows only 6 to 10 inches tall, spreading about a foot wide. It is evergreen, but last year's fronds die off in spring after the new ones have appeared.

Top: This is the hybrid tea rose 'Olympiad', a disease-resistant variety that doesn't require regular spraying. Bottom: The pink rose is 'Maiden's Blush', a centuries-old heritage variety. The red rose is 'Dortmund', which can be trained as a climber or as a large shrub.

ROSES

Modern Roses

Hybrid teas
***Rosa* hybrids**

Modern roses have a reputation for disease susceptibility. This is generally true, but in recent years All-America Rose Selections (AARS) has favored releases with superior disease resistance. Hybrid tea roses all need regular winter pruning, so they can't really be called easy; but see the box at right for some of the easiest. Although they still need pruning, they are more disease- and insect-resistant than most modern roses, so regular spraying is not necessary.

Easy Modern Roses

All are hybrid teas.

Variety	Color
'Broadway'	Yellow
'Confidence'	Pink
'Lady Rose'	Red
'Madame Violet'	Lavender
'Marijke Koopman'	Pink
'Olympiad'	Red
'Precious Platinum'	Deep red
'Pristine'	Ivory
'White Success'	White

Old Garden Rose

Alba rose, 'Maiden's Blush'
***Rosa* × *alba* 'Incarnata'**

Many antique roses are easy to care for. *R. alba* cultivars tolerate some shade and are very winter hardy, typically to –30° F. Their shortcoming is that they bloom just once, usually in June, so they are best in a wild garden or at the back of a bed of lower shrubs, where you won't mind that all they can offer most of the year is foliage. The flowers are worth the wait.

'Maiden's Blush' is a legendary rose more than two centuries old. It bears very fragrant double, blushed white flowers on a generous shrub as tall as 8 feet.

Miniature Roses

Miniature rose
***Rosa* species and hybrids**

In spite of their delicate appearance, miniature roses are much easier to grow than their full-sized cousins. They need no pruning, and many are resistant to diseases, so they need no spraying either.

No bigger than 2 feet tall, miniature roses are smaller versions of regular roses in every way—leaf, stem, and flower. They are perfect along formal pathways, at the front of the border, and in containers indoors or out, but they can easily look out of place amid wildflowers or large perennials. Their winter hardiness is more difficult to assess than that of larger roses, so if you plan to leave them outdoors anywhere from Zone 6 north, play it safe and cover

Easy Miniature Roses

All minis are easy, but these are the easiest.

Variety	Color
'Beauty Secret'	Red
'Child's Play'	White and pink
'Cinderella'	White and pink
'Gourmet Popcorn'	White
'Lavender Lace'	Lavender
'Mary Marshall'	Orange
'Over the Rainbow'	Red and yellow
'Rise 'n Shine'	Yellow
'Winsome'	Plum

them heavily with mulch in fall. See the box above for some disease-resistant varieties.

Climbing Rose

Climbing rose
***Rosa* × *kordesii* 'Dortmund'**

Climbing roses come in many varieties, single and double, in a wide range of colors. 'Dortmund', a German variety, bears a huge crop of slightly fragrant, single red flowers followed by orange fruit, called hips. It is dependable even in neglected situations as far north as Zone 6. The foliage is healthy and glossy, suggestive of holly. Where winters are colder, a good choice is the fragrant pink 'William Baffin', hardy to Zone 2, capable of growing 8 to 9 feet tall.

In places where climbing roses are marginally hardy, they should be grown near a south- or east-facing wall. They should be pruned back in late fall to about a foot from the ground, or pulled down, laid on the ground, and mulched.

English Rose

English rose
***Rosa* × 'Graham Thomas'**

The English roses are a twentieth-century series produced by English gardener David Austin. His varieties resemble old-fashioned roses in appearance and fragrance but are longer in bloom and more disease-resistant. They are among the easiest roses to grow and make excellent landscape shrubs. Some are also quite winter hardy. There are all colors and sizes, but one of the most popular is the golden-yellow 'Graham Thomas'. The plant is bushy and upright, 5 to 6 feet tall, with glossy, dark green foliage. 'Graham Thomas' is hardy to Zone 5.

Top left: 'Child's Play' is a sweetly fragrant miniature rose. Top right: 'Blanc Double de Coubert' is capable of withstanding winters in the upper Midwest. Bottom: 'Graham Thomas' is an English rose, a new class developed to have the charm of old-fashioned roses, but to be long-blooming and easy to care for.

Very Hardy Roses

Rose
***Rosa* species and cultivars**

Breeding programs in the Midwest and Canada have produced scores of extremely hardy, disease-resistant cultivars that are beautiful and also flower recurrently. Some of the best are 'Therese Bugnet', with double, pale pink flowers on a 6-foot shrub, and the newer 'Morden Blush', which is light pink, 2½ feet tall, and hardy to Zone 2.

Possibly even hardier, though briefer in bloom, are hybrids of *R. rugosa,* which can survive prairie winters without protection. Among the best are the fragrant, double, white 'Blanc Double de Coubert', 4 to 6 feet tall, or the single pink 'Frau Dagmar Hastrup', which produces red hips in abundance.

The Easiest Ground Covers

The "floor" of the garden is the background for plants. It connects the garden's many elements, but it can also be interesting and beautiful in its own right.

The most common ground cover is lawn, but like other ground covers, lawns need to be chosen carefully if you want an easy garden. The right choice can mean the difference between constant maintenance chores and almost none. Beyond lawn, there are many ground covers that are even easier to maintain, and each has its virtues. None of them needs mowing. Some flower, some turn color in autumn, some carpet so densely that no weeds can grow. All of them add interesting texture and appearance to the garden.

When you are choosing the right ground cover for your needs, you should consider not only your climate and the sun or shade where you want the plants to grow, but also how much traffic you expect in that area. Only lawn is strong enough to take a lot of traffic, and even lawn will be damaged if it is constantly walked upon. Pathways should be installed in the busiest places.

Ground covers such as lawn can grow on their own; others can be grown in association with neighboring plants, to lessen weeding chores and provide interest all season long. You will likely have to buy several plants in the beginning. They should be set close enough so that within one year they can touch. During the first year, the open areas between the spreading plants will need weeding.

Several plants described elsewhere in this book can be used as ground covers: daylilies, true geraniums, hosta, mint, ivy, and ostrich fern, and for temporary cover, the annuals impatiens, wax begonia, moss rose, and sweet alyssum.

Fine fescue is adapted to dry, shady forests. It is likely to make the most successful lawn under large trees, where other lawn grasses die out and leave bare spots.

LAWNS

Gardeners seldom think twice about including lawns as part of their landscape. You should think about whether you really want a lawn in preference to another ground cover, and if you do want lawn, you should realize that it consists of plants that need as much planning as your perennials and shrubs. Poorly chosen grass varieties for your lawn can be the source of endless work and frustration.

Of course, no lawn is maintenance free, unless you want a meadow of grasses allowed to grow tall and windblown, which is not a realistic option for most gardeners. To minimize mowing chores, the mower should be on its highest setting. This allows the grass to develop a strong root system, and means you do not need to mow as often to keep it looking neat. Taller grasses are also better able to compete with weeds than grasses clipped short.

No lawn grasses thrive in deep shade. The grasses best suited to shade are the fine fescues, but even they need some sun, such as the dappled light under a shade tree with a high canopy or the partial light by a wall facing east or west. It may be necessary to prune tree branches in areas where you want to grow lawn. Shallow tree roots will also compete with lawn. Under trees, the grass should be cut no shorter than an inch, so that it can establish a strong root system.

The recommended treatment for areas that are constantly shady is a mulch of stones or bark chips, or one of the shade-tolerant ground covers listed in this chapter.

Grown as an unmowed ground cover rather than a lawn, fine fescue forms an inviting and interesting pattern. It also makes the best lawn in shady areas, as can be seen in the photo on pages 62 and 63.

LAWN GRASSES

Grass for Shade in the North

Fine fescue
***Festuca* species and cultivars**

No lawn grass will tolerate deep shade, but fine fescues—grasses adapted to areas with cool summers—are the best choice for the dappled shade under a deciduous tree or the partial shade of an eastern or western exposure. In almost all soil conditions except constant wetness, these grasses will produce a good turf from seed in the first year. They require less water and fertilizer than bluegrasses, and are appreciated for the lowest maintenance requirements of all the popular northern grasses. In good soil, they can make a beautiful green lawn with fertilization as infrequent as once every 3 years. What they cannot tolerate are hot summer nights, which leave them vulnerable to disease.

Fescues make dense tufts of glossy, upright leaves. They are often mixed with bluegrass for a disease-resistant lawn adapted to full sun or partial shade. Their seed germinates quickly, so fescues are a good choice for reseeding thin areas in your lawn. The seed need only be sprinkled on the surface and watered thoroughly.

Fine fescues are seldom planted on their own except in areas that will be mowed infrequently. There, allowed to grow tall and windblown, they are one of the more ornamental grasses.

Grass for Shade in the South

Zoysiagrass
***Zoysia* spp.**

A green lawn where summers are hot is the promise of zoysiagrass. It will not tolerate the range of conditions that bermudagrass will, but in the warm gardens where it grows well—in the South, coastal New England, and the Pacific Northwest—zoysiagrass can produce an easy-care, luxurious lawn. It turns straw colored in winter in the South and with the first frost in the North. Like other shade-tolerant grasses, zoysiagrass needs some sun.

Zoysiagrass is planted from sod or roots and needs time to become established. It can take 2 years to grow into a thick, even carpet, and during that time it must be carefully weeded and watered. Once established, it is difficult to eliminate should you want to claim the space to grow something else.

Grass for Sun in the North

Kentucky bluegrass
Poa pratensis

The most popular of all lawn grasses for low temperatures, Kentucky bluegrass is adapted to all kinds of soils except those that are very wet, sandy, or acid. It is strong and tough, spreading slowly by rhizomes, seed, and tillers, which are plantlets on the stems. Grown from seed, Kentucky bluegrass requires 3 years to become fully mature and traffic resistant. When seeded, it should be mixed with a quicker growing variety, such as perennial ryegrass, to fill in until the bluegrass takes over.

Kentucky bluegrass grows rapidly in cool, moist weather but suffers in heat and drought. It browns, although it will turn green again rapidly as the weather cools. Warm, humid nights are the worst weather for Kentucky bluegrass. It may become diseased and may disappear after the first year in a climate that is too mild.

Kentucky bluegrass lawns do best when mowed as high as 2 inches, and their clippings make the best mulch of all the popular lawn grasses.

There are more than 20 varieties, so it is usually best to buy a mixture of varieties with different attributes. Look for the adjective *improved* on the label. A few types, such as 'A-34' and 'Nugget', will take some shade.

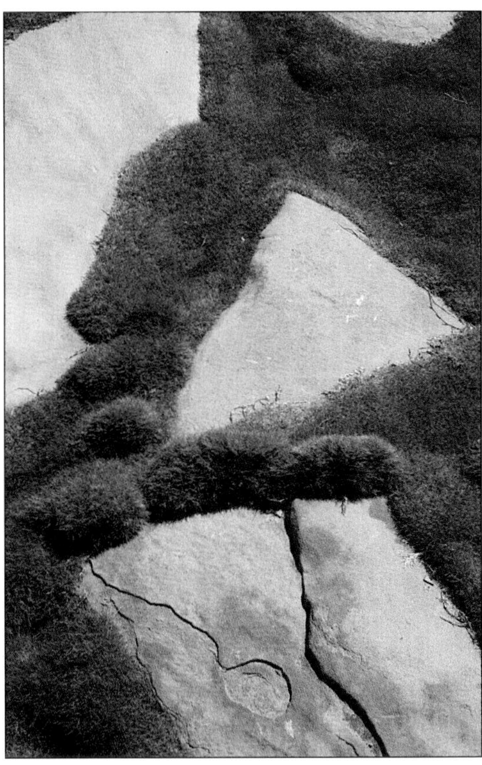

Top: Kentucky bluegrass is the most popular lawn grass for northern areas. Once established, it is tough and vigorous, making a dense turf that crowds out weeds. Bottom: Zoysiagrass forms interesting, low, mounded patterns, almost like a lawn, if not mowed. If mowed, it can have putting-green smoothness.

Grass for Sun in the South

Bermudagrass
Cynodon dactylon

A poor choice for northern gardens, bermudagrass becomes discolored at the first touch of frost, which leaves unattractive brown patches. But south of Washington, D.C., and especially in the sandy soils of the Atlantic coast and the Southwest, bermudagrass is the favorite lawn. It is suited to sun only, so if you have both sun and shade, bermudagrass should be blended with a shade-tolerant species such as Highland bentgrass.

Bermudagrass is a stoloniferous, or running, type of grass that can be propagated only from plants or sod, not seed. It can become a weed, so, where content, it should be enclosed to control its invasive nature. A pathway, curb, or raised bed can keep the runners from invading flower borders, vegetable gardens, and other open areas.

Common bermudagrass makes excellent lawns in the South if mowed closely. Named hybrid forms are also available; many of these are finer and denser than the common species.

In order to prevent matting, bermudagrass should be mowed very low. It is the only common type of lawn grass that should be maintained no higher than ½ inch. This grass turns brown in fall, as soon as temperatures drop into the 50° F range, and remains dormant until spring.

Grass Companions

White Dutch clover
Trifolium repens

Once favored for lawns, white Dutch clover fell into disregard when it did not fill the exacting requirements of the perfect, grass-only lawn. It is gaining popularity again because it needs no mowing, withstands heavy traffic, is tolerant of sun or partial shade, and, if allowed, produces fragrant white flowers in summer. Because clover fixes its own supply of nitrogen, it needs no fertilizer and can also help the growth of neighboring grasses. The flowers attract bees, which you may or may not find desirable.

Best suited to cool climates, white Dutch clover should be seeded into a lawn of Kentucky bluegrass or fine fescue. If you already have such a lawn, scatter clover seed over it in early spring, about a month before the first mowing, at the rate of a pound of seed for 2,000 square feet of lawn.

HERBACEOUS GROUND COVERS

Very Low Ground Cover for Sun

Mother-of-thyme
Thymus serpyllum* spp. *serpyllum

Tiny leaves that are aromatic when crushed are the distinguishing feature of this sturdy relative of culinary thyme. Mother-of-thyme will not take much foot traffic, but it is pretty and tough enough to serve as decorative filler between the paving stones of a pathway or patio in sun. It grows just a couple of inches high, so there is no danger of tripping unwary pedestrians. In summer it blooms rosy purple. It is hardy to Zone 5.

Mother-of-thyme can be sown from seed or grown from plants. It will need to be watered and weeded for the first couple of years, but after that it is drought resistant and strong enough to compete with weeds.

Mother-of-thyme is well suited to an association with small spring bulbs such as crocus, grape hyacinth, and early tulips.

Very Low Ground Cover With Yellow Flowers for Sun

Stonecrop
Sedum kamtschaticum

This succulent plant, also called sedum, comes in many sizes, flower colors, and leaf shapes. All are very drought resistant and able to put up with a summer of hot sun and dry weather without much watering. They will plump up quickly when rain does fall. Stonecrop does not tolerate much foot traffic. Some types are very invasive, but all are excellent in low pots that will contain them.

Restrained species such as *S. kamtschaticum* can also be used to front sandy borders or to edge a path or patio. It grows 2 to 4 inches high. The late-summer flowers are bright yellow-orange and the foliage is scalloped and evergreen.

Plants spread by rooting from the stems, so they are easily divided anytime of year. They should be watered a couple of times during the first week, to help them become established. This species is hardy to Zone 3.

Low Ground Cover With Blue Flowers for Shade

Bugleweed
Ajuga reptans

In spring, bugleweed produces spikes of flowers, usually blue, above ground-hugging, 6-inch-high rosettes of green, bronze, or variegated foliage, which is evergreen in warm areas. This lovely plant spreads by runners, so it fills its allotted space nicely. It may try to move beyond it, but in flower beds it can be pulled out easily, and mowing will control its advance into a neighboring lawn.

Bugleweed tolerates full sun but is most valued for its ability to grow in dappled shade under deciduous trees. It can take some foot traffic, so it is a good choice for places under trees where people sometimes walk. The most attractive types of bugleweed, though slower to spread than the species, are the purple-and-gold–leafed cultivar 'Multicolor', the green-and-cream–leafed 'Variegata', or 'Burgundy Glow', also called 'Burgundy Lace', which has purple leaves. Bugleweed is hardy to Zone 5.

Low Ground Cover With Gray Foliage for Sun

Snow-in-summer
Cerastium tomentosum

Named for the whitish appearance of the foliage and the summer flowers, snow-in-summer is a delightful ground cover where its rapid spread and dependability are appreciated. But its invasive nature means that in the good soil and cool climates it loves it should be kept away from delicate plants such as alpines and small bulbs. Confined by pavement or lawn, it can demonstrate its many benefits. It hugs the ground, grows about 9 inches tall, and is extremely winter hardy, to Zone 2. It is ideal as a no-maintenance substitute for lawn at the sunny front of a house, extending from foundation planting to sidewalk. This is also a good choice for permanent containers, where it will drape becomingly over the edges.

In cool climates, snow-in-summer should be planted in full sun in well-drained soil, even in sand. In warmer places it should be in partial shade, but it may not survive where summers are hot and humid. It is not dependable south

Top: Mother-of-thyme releases a strong, herby scent when trodden upon, making it ideal between steppingstones or beside paths, where its fragrance can be appreciated.
Center: Sedums come in a wide array of forms, sizes, and colors. This one, Sedum kamtschaticum, *is more restrained than many and less likely to invade its neighbors.*
Bottom: Bugleweed makes a splendid display of purple flower spikes in the spring. Its green leaves turn bronze in cold weather.

Top left: Snow-in-summer, with its gray leaves and white summer flowers, makes a cool contrast to brightly colored flowers.
Top right: Deadnettle 'Beacon's Silver' brightens shady areas with its white foliage.
Bottom left: Barrenwort is one of the most useful ground covers for shady areas in the North. This is 'Rose Queen'.
Bottom right: Lamb's-ears has softly downy leaves, giving it a teddy-bear quality that children love.

of Zone 7. Snow-in-summer spreads by rooting stems that are easy to move elsewhere if you want transplants for other parts of the garden.

Low Ground Cover With Gray Foliage for Shade

Deadnettle
***Lamium maculatum* 'Beacon's Silver'**

Although its name is unattractive, deadnettle is lovely, producing a ground-hugging cover of pointed, mintlike leaves on 8- to 12-inch stems in shade or in some sun. The small flowers of the cultivar 'Beacon's Silver' are purple. The leaves are silver, edged with green.

Deadnettle will put up with poor soil as long as it is well drained. It also tolerates drought but it dislikes hot weather. Give it partial to full shade. In heat, it tends to thin out, so it may fail from Zone 8 south, but in marginal areas it will grow thicker again when temperatures fall. In Zone 4 or colder, plants may die back after a severe winter.

Medium Ground Cover for Deep Shade

Barrenwort
***Epimedium grandiflorum* 'Rose Queen'**

Especially valuable for the humus-rich, moisture-retentive, acid soils under trees, barrenwort grown in this situation forms a dense ground cover in Zones 5 to 8. The spring flowers are lovely, suggestive of columbine. The cultivar 'Rose Queen' has burgundy leaves and pink flowers. It grows almost a foot tall.

Plants should be set a foot apart in deep shade. Barrenwort will burn in sun unless the soil is kept constantly moist. It is not invasive, spreading slowly by rhizomes.

Medium Ground Cover for Shade

Brunnera
***Brunnera macrophylla* 'Langtrees'**

Happiest in the dappled shade under trees in moisture-retentive soil, brunnera produces a clump of leaves that grow directly from the soil and look fresh and showy all season. Blue flowers bloom in spring and may last more than a month; then the plant resembles a giant forget-me-not. The leaves, which grow more than a foot tall, die back to the ground in winter. In hot regions, brunnera needs a consistent moisture supply; an area near a stream or pond is ideal. 'Langtrees' is the most drought tolerant cultivar, but it too suffers in prolonged periods of dryness. It is not worth trying in a dry garden because the leaves, the plant's showiest feature, will wilt and turn brown.

Brunnera, also known as *Anchusa myosotidiflora,* grows best in Zones 4 to 7.

Medium Ground Cover for Dry Sun

Lamb's-ears
***Stachys byzantina* 'Silver Carpet'**

The soft, furry foliage of lamb's-ears, which gives this plant its name, is an endearing feature that tempts passersby to touch it. In full or partial sun, even in very dry, poor soil, lamb's-ears will form a widening clump of foliage. In fact, lamb's-ears does poorly in rich soil and will fail in a wet area.

'Silver Carpet' is a nonflowering cultivar, which is an advantage for this plant. The flowers of other cultivars are small, purple, and not particularly showy. To keep plants looking neat, the flower spikes should be removed after blooming.

Lamb's-ears is good for a south-facing slope by a driveway or path. It looks best in the company of other dry-soil plants such as stonecrop and mother-of-thyme. Despite its softness, it can put up with a bit of foot traffic. It is hardy to about Zone 5.

WOODY GROUND COVERS

Very Low Ground Cover for Sun

Prostrate broom
***Genista pilosa* 'Vancouver Gold'**

Although it has a somewhat unkempt appearance, due to its slightly hairy leaves and tangle of gray-green branches, prostrate broom is valued as an easy ground cover because of its considerable tolerance of heat and sun. In a sunny place, it will form a dense, evergreen cover about 6 inches high. The golden-yellow flowers bloom in clusters in late summer, lasting as long as 6 weeks.

Top: Prostrate broom 'Vancouver Gold' is covered with sun-yellow blossoms for several weeks in the summer. Bottom: Bearberry 'Vancouver Jade' is a low-growing manzanita. It thrives with little care almost anyplace but in clay soil.

This is a good cover under a border of sunny shrubs, and it can also spill over a wall or grow in containers. Each plant spreads slowly to more than 3 feet wide, rooting as it spreads. It is hardy from Zone 6 south.

Low Ground Cover for the North

Bearberry
***Arctostaphyllos uva-ursi* 'Vancouver Jade'**

This northern native is evergreen in gardens where winter temperatures dip as low as –50° F, Zone 1. Bearberry demands only well-drained, acid soil, but otherwise will put up with sun or shade, poor soil or rich, forming a beautiful, small-leafed carpet 6 to 12 inches deep. It tolerates considerable drought or wetness. The foliage is green in summer, bronze purple in winter.

Fragrant pink flowers are followed by bright red berries. Growth can be slow at first, so plants should be set about 3 feet apart, but eventually a plant can cover a 5-foot circle. The cultivar 'Vancouver Jade' is especially vigorous and disease resistant.

Top: Pachysandra forms a lovely deep carpet in perfect harmony with trees and stone. Bottom: The bright blue flowers of periwinkle in the spring are a bonus; it is grown primarily for its glossy, dark green leaves.

Low Ground Cover for Shade

Pachysandra
Pachysandra terminalis

This is the ground cover of choice for most shady gardens. Also called Japanese spurge, pachysandra is valuable for dry, shady slopes where little else will grow well. In sun, the foliage burns. The plant grows about a foot tall, creating a solid mat of green foliage whose starlike growth pattern creates textural interest under trees, next to fences, and in other places where there is no foot traffic. Tall tulips and daffodils can be interplanted with pachysandra and will bloom through it. Pachysandra spreads by runners; it is best controlled by a wall or pavement or a mowed lawn.

Pachysandra prefers moist, acid soil, but it will tolerate drought. It forms a dense cover to Zone 5, a more sparse cover in areas as cold as Zone 3.

Low Ground Cover With Blue Flowers for Shade

Periwinkle
Vinca minor

Also called myrtle, periwinkle has all the virtues of the best garden plants: attractive, evergreen foliage; winter hardiness; lovely flowers; and virtually no need for a gardener's care. It also makes a very good ground cover, spreading by rooting stems. Starting out as a ground hugger, it develops into dense mounds of foliage and stems that can eventually become more than a foot deep. It is well suited to covering the ground beneath shrubs and deciduous trees, or to blanketing an area that has almost no foot traffic. It can discourage weeds in a perennial bed, and large bulbs such as tulips and daffodils planted in the same area will bloom through it.

The usual form has flowers that are periwinkle blue, but you might also find white- and purple-flowered forms. Periwinkle is hardy to Zone 3.

Periwinkle can move beyond bounds if the stems are not pulled up where they are not wanted. It should not be grown next to very delicate or small plants.

Low Ground Cover With White Flowers for Shade

Sweet woodruff
Galium odoratum

White flowers shine like tiny stars from sweet woodruff, a beautiful little ground cover that is happiest under trees or roses or in other partly shaded places. It suffers in sun and drought. The finely divided foliage is delicate and attractive all season. Dried, it has a sweet aroma.

Plants hug the ground, the reclining stems reaching up to 6 to 12 inches. Sweet woodruff is hardy to Zone 4 and spreads strongly but is easy to root out. It can't tolerate much foot traffic.

Medium Ground Cover for Sun

Rock cotoneaster
***Cotoneaster horizontalis* 'Deneen'**

Generally considered a shrub, cotoneaster has wide-spreading forms that make excellent ground covers, especially because of the bright red berries. The plant is deciduous, with good fall color.

The cultivar 'Deneen' is leafy from the ground to a height of 2 to 3 feet. It spreads outward, rooting wherever its branches touch the ground. It must be tended and weeded while it is becoming established but will eventually require little care. 'Deneen' can be used to cover a difficult slope, or planted around rocks or behind prostrate junipers. It should be given full sun or a little shade. This cultivar is hardy to Zone 6.

Tall Ground Cover for Sun

Pfitzer juniper
***Juniperus* × *media* 'Pfitzeriana'**

Used as ground covers, junipers are evergreen, sturdy, and attractive. They cast such dense shade that few weeds can grow through them. Probably the most popular garden juniper, the pfitzer assumes a bird's-nest shape, with its branches at a 45-degree angle, and eventually reaches at least 5 feet tall and 10 feet wide. The foliage is blue-green. For use as a ground cover, the plants should be set approximately 3 feet apart.

'Pfitzeriana' is hardy to Zone 3, although it may suffer from winter damage in the northern part of its range. It is massive and stately if given plenty of space and dry, well-drained soil in full sun.

Junipers have prickly foliage, so they should not be planted where adjacent plants will be tended with your bare hands. Junipers are ideal on the sunny side of a group of mixed shrubs. This species is less troubled by juniper blight than *J. horizontalis*, another common ground cover.

Left: Sweet woodruff is used in Germany to flavor May wine. It is also a shade-tolerant ground cover.
Right: This cotoneaster, 'Deneen', stays low and spreading. It is covered with sprays of white flowers in the spring, and with red berries in fall.

The Easiest Vegetables and Herbs

Nothing can compare with vegetables and herbs fresh from your own garden. Here are the easiest, most self-reliant kinds.

Sun-ripened tomatoes, crunchy lettuce, fragrant basil, and pungent chives are just a few of the vegetables and herbs that are suited to the easy garden. Recognize, though, that "easy" does not mean no work. Most vegetables and some herbs are annuals, so they must be planted every year, either from seed or transplants. As with any newly planted thing, they require watering and weeding to grow. Some garden preparation is also required.

Vegetables always taste better and are healthier and more productive if they grow as quickly as possible. This means selecting a sunny spot—generally with at least six hours of sun a day—with good, well-drained soil, and giving them lots of fertilizer and water.

Vegetables are easiest to care for in permanent beds. Each bed should be narrow enough so you can reach the center without stepping in the bed—about 4 feet for most people—and less than 10 feet long. Paths between the beds should be 18 inches wide, or wide enough for your garden cart. When you begin a new garden, spread 2 to 4 inches of compost, manure, or some other organic soil amendment over the soil and turn it under. Raised beds enclosed by wood or masonry walls are easiest to care for, but a mounded bed offers many of the same benefits. Cover the paths between the beds with sawdust, gravel, or some other mulch to prevent weeds.

In dry-summer areas, you can save time by installing a watering system. The most effective consists of a porous hose that can be laid on the surface of the soil or buried an inch deep. You probably need two

Vegetables, like any annuals, must be planted and harvested each year, making them less "easy" than completely care-free perennials. However, some vegetables require less attention than others.

lengths of leaky hose for each bed. You can attach a garden hose to each leaky hose or install a permanent system of black plastic hose leading to each bed. A permanent system can be fitted with an inexpensive battery-powered timer that will turn the water on and off automatically.

After you plant each bed, cover the soil with 3 inches of mulch. Fallen leaves, hay, or grass clippings not treated with herbicides are best, but any organic material that will decompose in the soil will work. The mulch will keep the surface of the soil cool, conserve water, and prevent weeds from growing. If any weeds sprout, cover them with a handful of mulch.

Turn under the mulch as a soil amendment before planting the next crop in that spot. After a couple of years, the soil will be as dark and soft as chocolate cake crumbs.

Sprinkle on all-purpose or vegetable fertilizer once or twice as the plants grow.

The vegetables described in this chapter are the easiest ones to grow. In most cases, one or more varieties are recommended; these are usually more resistant to disease or easier to grow than other varieties. In some cases, any variety of the vegetable is easy to grow.

Several vegetables, such as celery, head lettuce, and cauliflower, are missing from this list because they are not easy—they need ideal soil and climate conditions or a high level of skill and expertise to grow well.

None of the varieties recommended are rare, but no seed catalog carries all of them, so you may have to investigate several catalogs to find a recommended variety.

THE EASIEST VEGETABLES

Bean

For the North: 'Provider'
For the South: 'Derby'

Both of these varieties are bush snap beans. 'Provider' is an early bean that will germinate in cool soil. 'Derby' is an adaptable bean that will set a bountiful crop in spite of adverse conditions. Bush beans do not need any support or special care.

Around the time of the last spring frost, sow bean seeds an inch deep and 3 inches apart. Plant about half your crop at once, then plant a few more seeds every 2 weeks until mid-summer. The later plantings will continue the harvest until frost. Pick beans when they are young and tender, just as the beans begin to swell in the pod. Picking every 3 days prolongs the harvest.

Chard

Any variety

Chard might be the easiest and most rewarding vegetable in the garden. It can be started from seed or plants, will produce leaves in the hottest weather, and even survives light frosts, growing into the fall. It is adaptable to many soils and climates and is rarely bothered by pests or diseases. Unlike many vegetables, it can even take a bit of shade.

Left: These are 'Derby' bush snap beans. Pick all the beans when they're ready; if some are allowed to ripen on the bush, the plant stops bearing.

Right: Swiss chard is easy and sure, and it is also very productive; you can harvest many pounds of leaves from just a few square feet over the course of a growing season.

Start from 4 to 8 plants for a small family. Once the plants are about a foot high, harvest one or two leaves at a time from each plant, taking the oldest, or outside, leaves. Plants will continue to grow until killed by frost. If you live in an area with mild winters, you can harvest chard all winter long, until the plant goes to seed, or bolts, in spring.

Chard is an easy substitute for spinach. The large leaves are decorative enough that chard can be planted between flowers in an annual or a perennial bed. The red-leafed types are especially beautiful.

Top: The mild 'Anaheim' chile pepper makes a lively show in the vegetable garden. Bottom: Cucumbers may be fussy about weather and subject to diseases. 'Little Leaf' is easier to grow than most.

Chile

Mild hot: 'Anaheim'
Medium hot: 'Hungarian Hot Wax'
Really hot: 'Habanero'

Chiles are like smaller, hotter versions of bell peppers. They are easier to grow than bell peppers, demanding less water, less fertilizer, and less space. In areas with hot summers, they are easy to grow. In areas with shorter, cooler summers, grow 'Hungarian Hot Wax'; this medium-hot variety sets fruit in cooler weather. 'Anaheim' is a mild hot pepper, the variety commonly used for *chiles rellenos.* 'Habanero' is *extremely* hot; handle it with care! In Jamaica, where it is used to make the hot jerk sauce, it is called 'Scotch Bonnet'.

Chiles are decorative plants, with their woody stems, green leaves, and small white flowers followed by colorful fruits, so they are well suited to growing in a bed of flowers or in a container on a porch or patio. Three plants will fit in a pot a foot wide and deep.

The chiles can be picked any time after they are full grown. If they are left on the plant, they will change color, gradually ripening to red, yellow, or orange.

Cucumber

'Little Leaf'

This bush-type cucumber is exceptionally dependable; it adapts well to a variety of soil and climate conditions and is resistant to several diseases and insects.

Cucumbers are hot-weather plants and produce best in the heat of summer. Wait until all danger of spring frost is past and nights are warm to plant them. Select a sunny place with well-drained soil. Water deeply once a week if the weather is dry.

Harvest cucumbers regularly to prolong fruit production. Cucumbers will not grow or set fruit in cold weather and will die with the first frost.

Lettuce

Loose-leaf: 'Black-seeded Simpson'
Romaine: 'Parris Island 318'

Head lettuce may be what you usually buy, but leaf lettuce is easier to grow and is often better tasting. 'Black-seeded Simpson' is lime green

Top: 'Black Seeded Simpson' is an old-fashioned variety that has been around since the turn of the century, and is still a favorite for its dependability and mild flavor. Bottom: Snap peas, such as 'Super Sugar Mel', are a relatively new vegetable. Because the pod is as sweet and edible as the seeds inside, they produce more food than garden peas.

in color and has a sweet, delicate flavor. It is early, large, and vigorous, and tolerates heat better than most lettuces. 'Parris Island 318' is an improved variety of 'Parris Island'.

Lettuce does best in cool, moist soil and fairly cool weather. Sow seed in early spring, as soon as the soil can be worked. Sprinkle the seeds on a weeded patch of soil and rake them in lightly, then water until they sprout. Thin to 6 inches part. If you buy young plants, wait until around the time of the last spring frost to set them 6 inches apart in fertile, well-drained soil. If your summers are hot, plant lettuce in partial shade; where the season is cool or short, give it full sun.

Although lettuce can be harvested a few leaves at a time, like chard, it is sweeter and more tender if whole heads are harvested while they are still young. As you harvest each head, place another plant or seed in its space. Continue to sow or set out transplants until midsummer. Where winters are mild, sow lettuce in late summer for winter and spring harvests.

Melon

For the North: 'Earligold F1'
For elsewhere: 'Pulsar'

Melons are generally not easy to grow well. They require long, hot summers with warm nights to be sweet and tasty, and they are subject to several diseases. The two cantaloupe listed here are more likely than other melons to be successful. Watermelons and some other types of melons are generally more difficult than cantaloupe.

Give melons full sun and rich, well-drained soil. If you don't live in an area with long summers and warm summer nights, start the plants indoors in peat pots. A couple of weeks before setting them out (when the nights have become warm), cover the ground where they will grow with clear plastic. This will warm the soil. Cut holes in the plastic to plant the melons. If the weather becomes very hot before the melons grow enough to cover the plastic with leaves, spread some mulch over the plastic to shade the soil; otherwise the soil can become so hot it will kill the melons. After you set out the melons, cover them with a floating row cover. This lightweight spun fabric can be purchased in garden centers. It is so light it does not need to be propped up; it will lift as the plants grow. Weight the edges with soil to keep the row cover from blowing away. Remove the cover when the weather turns hot.

If possible, let the melon vines grow across an asphalt driveway. The black asphalt collects heat during the day, which keeps the vines warm at night.

Pea

Garden pea: 'Knight'
Snap pea: 'Super Sugar Mel'

Peas are cool-weather crops. Plant the seeds 2 inches apart as soon as the soil can be worked in spring. In mild-winter areas, plant in early

October for an early spring crop. Both the varieties recommended here are short and can be grown without support, but they are easier to pick if grown on brush or a low fence, 2 to 3 feet high.

Pick peas every couple of days. Garden peas are ready to pick when the peas have filled the pods but the pods are still shiny and smooth. Snap peas—which are eaten whole, pods and all—are ready as soon as the pods reach full size and the peas have begun to swell. Snap peas have strings; as you pick each one, twist it back to leave the string on the plant.

The sugar in peas begins turning to starch as soon as they are picked, so plan to pick them just before dinner. Newly picked peas are so sweet that some gardeners don't cook them but just eat them raw.

Peas stop producing when the weather becomes hot. Then their vines can be pulled out and the area can be filled with a late crop such as chard or beans.

Pumpkins are fun to grow. Children love to watch the fruit swell and turn orange just in time for Halloween. This is 'Howden'.

Potato

'Butte'

Potatoes need a lot of space. For each hill, allow a circle 2 feet across. About a month before the last spring frost, cut seed potatoes into egg-sized chunks, each of which should have at least one eye. Plant these chunks eye upward, 6 inches deep in any weed-free soil. New growth may be killed by late spring frosts but will regrow. Potatoes can also be grown in half barrels, 6 chunks per half barrel.

Potatoes are ready to harvest when the tops die back to the ground in late summer. With a garden fork dig them carefully to avoid piercing the potatoes; start from the outside of the plant and work toward the center.

To grow "new," or immature, potatoes, place the chunk of seed potato on the surface of the ground and cover it with 6 inches of mulch. The plant will grow through the mulch without difficulty. The roots will grow into the soil; the potatoes will form between the mulch and the soil. You can harvest the immature potatoes by feeling carefully under the mulch. Pick those that are the size of a golf ball. Of course, if you pick many when they are this size, you'll have fewer mature potatoes. Some gardeners reserve a few plants just for new potatoes.

Pumpkin and Winter Squash

Pumpkin: 'Howden'
Winter squash: 'Table King'

'Howden' is a large pumpkin but not one of the giants. It is a good size for jack-o'-lanterns. Pumpkins like to ramble and thus need a lot of space. If you have a place where the vines can grow unimpeded, this is an easy crop that pays dividends in terms of fall pies and Halloween decorations. Plant in groups of 6 seeds, with the groups 6 feet apart. Thin to 3 plants in each group when the plants are 2 inches high.

'Table King' is a bush acorn squash. It takes up only a few square feet, so it is easier to grow in a small area than vining squashes and pumpkins. Plant seeds in groups of three, with the groups 3 feet apart. Thin each group to a single plant when the plants are 2 inches high.

Winter squashes and pumpkins are most easily grown from seed, not plants. Sow seeds ½ inch deep in fertile soil after the ground has warmed. If your summer is short, start seeds indoors; use clear plastic when transplanting to the garden, as described for melon on the opposite page. If frost comes before the pumpkins mature, pick any that are beginning to color and put them in a sheltered, airy place, such as on a porch, to finish ripening for Halloween.

Top left: Radishes are one of the easiest vegetables to grow. Shown here are 'Sparkler', with white tips, and 'Champion'.
Top right: 'Gold Rush' is a golden zucchini. It has a mild, sweet flavor.
Bottom: 'Viva Italia', a paste tomato, is thick and meaty with little juice, making it a good salad tomato as well.

Radish

Any variety

Radishes are the earliest-maturing vegetable and one of the easiest to grow. Globe radishes, the small type often used as a garnish, are perfect to grow between other vegetables in the garden or in containers. Sow seed ¼ inch deep in early spring; thin to 3 inches apart. Radishes will sprout in 5 days and be ready to harvest in 3 weeks. Do not leave them in the ground much longer or they will become hot and woody. Continue sowing radish seed every week in spring as long as the weather remains cool.

There are red, white, and bicolored types of globe radishes, all of which are better grown from seed, not plants. The standard red globe is 'Cherry Belle'. Whites include 'Burpee White' and 'Snowbelle'. The usual bicolor is 'French Breakfast', which is cylindrical.

Because radishes grow so easily and quickly, they are the perfect crop for children to plant.

Summer Squash

'Gold Rush F1'

Most zucchini varieties are easy to grow—so easy, in fact, that most gardeners have too many zucchini. Two plants are enough for a small family. 'Gold Rush F1', a golden zucchini, is flavorful and more attractive than many other varieties.

There are other types of summer squash as well: straightneck, crookneck, pattypan (or scallop), and vegetable marrow. All are easy and all are best picked when young and tender.

Each summer squash plant needs a space about 4 feet across. Or one plant will fill a half barrel and will drape over the sides. Container-grown plants must be watered as frequently as once a day in hot, dry weather.

Squashes are warm-weather vegetables. Plant them after the soil has warmed in spring. Sow them in groups of 3 seeds, and thin out all but the strongest one when they are a couple of inches high. The first harvest should take place 4 weeks after planting.

Tomato Disease Resistance

Resistance to tomato disease is shown by the string of capital letters after the name. Each letter stands for a disease to which that tomato is resistant. Some older varieties are resistant to one or more of these diseases but have not been formally rated and renamed. The lack of letters doesn't necessarily mean that a variety is susceptible to all diseases.

V = Verticillium wilt

F = Fusarium wilt. Two *F*s indicate resistance to both of the two races of fusarium.

N = Nematodes. These aren't a disease but a tiny, wormlike creature that attacks the roots.

T = Tobacco mosaic virus

A = Alternaria leaf spot

Basil is an annual that can be easily grown from seed scattered in a corner of the vegetable garden. Harvest it fresh for flavoring salads, stews, or tomato dishes. 'Genovese' is pictured.

Tomato

Short seasons and cool summers: 'Oregon Spring'
Large tomato: 'Big Beef VFFNTA Hybrid'
Small tomato: 'Enchantment VFFN Hybrid'
Paste tomato: 'Viva Italia VFFNA Hybrid'

Several tomatoes are suggested, each one the easiest in its class.

'Oregon Spring' and 'Viva Italia' are *determinate* tomatoes. They make a small bush; set a single, large crop of fruit; and then are done for the season. They need no staking or support. The other two types are *indeterminate* tomatoes—vine types that keep growing and making fruit until killed by frost. For the easiest picking and care, grow them in a tomato cage to keep them upright and off the ground. Tomato cages can be purchased in garden centers.

Several types of tomatoes are suited to containers. The vining cherry tomatoes, such as 'Sweet 100', will drape over the sides of a half barrel. Bush-type cherry tomatoes such as 'Small Fry' produce their fruit on a plant that grows upright. Three plants will fit in a half barrel. There are also larger-fruiting plants suitable for containers, such as the 'Patio' or 'Husky' series.

Give tomatoes a place sheltered from strong winds, in deep, well-drained soil. If the weather is dry, water tomatoes deeply once a week.

THE EASIEST HERBS

Herbs are some of the easiest plants you can grow. The characteristic fragrances and flavors we prize come from natural pest deterrents found in the plants. Herbs also tend to be free of diseases. Many are from areas with hot, dry summers, and they develop their best flavor if grown without too much water or fertilizer.

There are annual, biennial, and perennial herbs. You can grow the perennials on their own in a special herb garden, but all of the following herbs are also decorative enough to grow between flowers, or to highlight in containers. The annual and biennial types can also be grown in a vegetable garden.

Although some of the herbs described here are available in several varieties, all varieties are easy to grow, so specific varieties aren't recommended. Most of the difference in varieties lies in appearance and flavor.

Annual and Biennial Herbs

Basil

No garden should be without this easy and attractive annual. A natural kitchen companion for tomatoes, it does best in much the same garden conditions: warmth, sun, and well-drained soil. Basil stops growing or fades in cool weather and dies after a frost.

Left: Two types of parsley are commonly grown. The plain "flat Italian" type has a strong, distinctive flavor; the crinkled type, such as the 'Dark Moss Curled' pictured here, is often selected for garnishes.
Right: Bay laurel can be grown as a container plant—leave it outside for the summer, then bring it indoors before freezing weather.

Basil can be grown in a patch or row in the vegetable garden, but it is also suited to a sunny spot in the flower garden as well as to containers and even window boxes. There are many types. Leaves may be green or burgundy. The fragrance may suggest lemon, camphor, cinnamon, licorice, or cloves; new types are being developed constantly. The leaves may be large or tiny. One small-leafed type, 'Spicy Globe', grows into a perfect 6-inch ball that is showy enough to edge a pathway.

Grow basil from purchased plants or from seed. Set plants 6 inches apart. Sow seed in rows or in clusters, an inch apart, gradually thinning the seedlings as they grow. Basil requires some watering if the weather is hot and dry but is otherwise self-reliant.

Dill

Sprinkle a few dill seeds on the ground around other plants or in their own patch, water them until they sprout, and you will have delicious dill leaves in late spring, followed in summer by umbels of dill seed to use in pickles and salads. Dill is an annual that does best in full sun, though it will tolerate some shade. It is not demanding about soil, but it docs grow taller in fertile soil.

Dill is a decorative plant. The cultivar 'Fernleaf' was developed especially for flower beds. Its flower stalks grow 18 inches tall, whereas the species can reach 2 feet. Leave some seed heads in the garden to self-sow.

Parsley

Although parsley is a biennial, which means that it produces seed in its second year, it is the foliage, not the seed, that gardeners want, so parsley is most valued in its first year and is grown as an annual. It does best in cool, moist ground, but it will tolerate some drought and can be grown in sun or shade. It is decorative enough to grow alone in a pot or to fill the spaces between perennial or annual flowers, contributing a cushion of bright green leaves all season long.

If you leave parsley in the garden over the winter, it will produce dill-like seed heads the following spring, then it will usually die. Self-sown seedlings can give you a supply of parsley from then on. Parsley does not transplant well except when tiny. If you buy plants, be careful not to disturb the roots when transplanting them. If it self-sows, move the seedlings early, or thin them instead of transplanting them.

Perennial Herbs

Bay laurel

Where it is hardy, bay laurel (*Umbellularia californica*) forms a glossy evergreen shrub of edible leaves that can take its place among flowers or other shrubs. It prefers well-drained soil and dappled or partial shade. The foliage burns in hot sun. Bay laurel becomes hardier with age and increasingly tolerant of light frost.

Top: Chives are as ornamental as they are tasty. Grow in a pot or clump by the kitchen door so you can step outside to harvest a sprinkling for the salad or baked potatoes.
Bottom: With its attractive flowers, lavender is as familiar in the perennial border as the herb garden. This is 'Munstead'.

It is hardy to Zone 7; in colder areas it can be grown as a houseplant, either wintered indoors or kept indoors all year.

Although bay laurel grows slowly, it can be pruned if it gets beyond bounds. This is a fairly drought resistant plant that needs deep watering only about every 2 weeks.

Chives and garlic chives

With their tubular, dark green leaves and lilac flowers, chives (*Allium schoenoprasum*) are decorative enough for an annual or a perennial flower bed. They are also attractive container plants. They will grow in dry or wet soil, sun or partial shade. Garlic chives (*A. tuberosum*) are just as easy. The garlic-flavored leaves are flattened, not tubular, and the flowers are white, on 30-inch stems. Add cut-up bits of garlic chive leaves to salads for a mild garlic flavor.

Chives can be grown from seed sown in early spring, though they are easiest from the small onion bulbs at the base of the plant. Where content, they form a gradually widening clump, and will self-sow as well. Seedlings are easy to remove if you wish. Chives are hardy to Zone 4 and evergreen in Zones 9 and 10. In colder regions of its range, chives die to the ground in winter and reappear in spring.

Harvest as needed by cutting a few leaves from the outside of the plant rather than by giving the entire plant a "haircut." Harvesting whole leaves keeps the attractive shape of the plant and allows it to bloom.

Lavender

Grown for fragrance rather than for cooking, lavender is an attractive garden plant, with small grayish leaves and, in summer, spires of light purple flowers from 1 to 3 feet tall depending upon the cultivar. Lavender needs well-drained soil and prefers sun or light shade. Some types are quite frost-tender, although a few, such as *Lavandula angustifolia* 'Munstead', are hardy in sheltered spots as far north as Zone 5. Snow cover or a protective layer of fir boughs helps lavender survive cold winters.

Lavender is most easily grown from plants, although the All-America Selection 'Lady' is easy to grow from seed sown in spring.

Lovage

Although not well known in America, lovage is esteemed in much of Europe, especially as a seasoning for soup. In flavor and appearance, lovage suggests both parsley and celery, two of its cousins. Lovage is much larger, however. Given good garden soil and sun, it grows 5 feet tall and forms a gradually widening clump. This is a domineering plant best grown where it will not shade other things. Grow it at the northern end of a vegetable garden or surrounded by shrubs. It is hardy to Zone 4. If you cannot find a plant, grow it from seed sown in fall.

Mint

Where mint is happy, it can be difficult to control, but where it is unhappy, it can be difficult to keep. Mint grows best in moist, fertile soil. Given that, mint will put up with sun or shade, heat or cold. In places ideal for mint, keep it in bounds by growing it in a restricted area, bordered by lawn and foundations, or by surrounding it with an underground barrier, such as the 5-gallon metal or plastic cans that hold nursery plants. Cut out the bottom of the can and bury it with 2 inches of rim exposed. Plant the mint in the can. Check every few weeks during the summer to be sure that the mint is not escaping its container by sending stems over the can to root outside.

Spearmint (*Mentha spicata*) is the classic mint with the best flavor for all-around use in the kitchen. It is hardy to Zone 4.

Good-quality mint must be started from a cutting, not seed. It forms a ground cover 2 to 3 feet tall, spreading outward by means of stems and underground rhizomes.

Rosemary

A tender perennial, rosemary is hardy only to Zone 7. In the South, it can grow into a 4-foot, aromatic shrub, ideal as a focal point in the

Top: Lovage shows its relationship to celery not only in the shape of its leaves, but also in its flavor.
Bottom: Spearmint can be a real problem if it's free to roam, but is completely carefree if planted in a contained area.

center of an herb garden. Farther north it remains so small and tidy that it makes a good container plant that can be wintered indoors. Set one plant in a pot 8 inches wide, in a mixture of compost and sand. Give it the sunniest window when it comes indoors, and water regularly, using room-temperature water.

In the garden, give rosemary well-drained soil, in full sun or a little shade. Rosemary needs a deep watering once a week.

Sage

Garden sage (*Salvia officinalis*), the type of sage used for cooking, has grayish leaves and woody stems about 18 inches tall. It delights in dry soil, warm summers, and full sun. If your garden is wet and shady, put sage in a pot of sandy soil in the brightest place you have.

Sage is perennial but tends to be short-lived, especially in the northernmost reaches of its range, Zone 4. To keep it going, divide it in spring after 2 to 3 years. Plants set in the ground should be spaced a foot apart.

Thyme

The thyme commonly sold for kitchen use is English thyme (*Thymus vulgaris*). Another variety, known as German winter thyme, is hardy to Zone 3. More tender types include caraway thyme (*T. herba-barona*) and lemon thyme (*T.* × *citriodorus*). All thymes are easy to grow, demanding only some sun, well-drained soil, and weeding until they form a ground cover dense enough to be self-weeding.

One species of thyme, *T. serpyllum* spp. *serpyllum,* is described in the chapter on ground covers. Other species, too, make acceptable ground covers, with woody stems and a reclining habit, though they are not as resistant to traffic as *T. serpyllum* spp. *serpyllum.*

Top left: Drought- and heat-resistant, rosemary is easy to grow in a container. Use sprigs to flavor chicken or lamb, and for a garnish.
Top right: Like the other herbs on this page, and many herbs in general, sage is native to the Mediterranean region, with its hot, dry summers. To develop the strongest flavor, grow it in full sun and don't water it too often.
Bottom: Thyme is one of the most useful herbs in the kitchen, being used to flavor many meat and vegetable dishes.

The Easiest Houseplants

The indoor garden can give pleasure the year around without demanding much time or trouble. Here are the houseplants to grow if you want beauty but not work.

The plants in this chapter are divided into three categories, based on their function in interior decor. Decide what kind of plant you want in a specific location, then look up that shape or type in the list that follows to determine the easiest plant to grow.

Foliage houseplants are valued for the appearance of the plant as a whole, rather than just for its flowers. Most foliage plants do have flowers, but either the flowers are insignificant and are not noticed, or the conditions are not suitable for flowering.

The climbing plants will energetically produce foliage that can be trained to climb and drape.

The easy flowering houseplants offer a choice of form, color, and blooming season. You can also consider wax begonia (page 46), impatiens (page 47), and geranium (page 45), all of which will grow happily in containers indoors.

Houseplants have the same requirements as outdoor plants—they need light, water, fertilizer, and a certain temperature range. Outdoors, nature provides most of their needs. Indoors, you are the provider.

Bird-of-paradise makes a dramatic houseplant, complementing modern interiors as well as Victorian ones. See page 89 for a description.

A light garden allows you to grow African violets in any unused corner of the house, bringing them out to ornament a table or windowsill when they are looking their best.

LIGHT

The houseplant requirement that calls for the most attention is light. Light provides plants with the energy they need for photosynthesis—the process by which plants manufacture food from water and from the carbon dioxide in the air. Light is more important for their survival than fertilizer is; without enough light, plants grow slowly or not at all, and finally starve to death.

In general, cactuses, succulents, and "outside" plants need full sun for several hours a day. Flowering houseplants won't tolerate direct sunlight, but they need bright light to bloom. Foliage plants need the least light.

South-facing windowsills have the brightest light. This might seem ideal for houseplants but in fact, since most houseplants are native to dense tropical forests, they aren't happy in full sun. Cactuses and succulents, however, will thrive in such a location. Miniature roses and some other "outdoor" plants also make satisfactory houseplants on a sunny windowsill. To adapt a full-sun location for plants that won't tolerate it, hang a lace curtain. It will let through enough light for the plants to thrive.

Windowsills that face north or are shaded by a tree are good habitats for most houseplants. The plants receive lots of light through the window but no direct sunlight.

Many houseplants can be raised anyplace in the house by training a light on them. Use track lights or any light that can be aimed to concentrate its beam.

Another way to keep houseplants healthy is to rotate two plants between bright and dim locations. For example, place an African violet on a table in a low-light location and another in a north-facing window where it will get plenty of light. Every two weeks, switch the plants.

If you cannot find a location with the right light level, or if you would like to raise additional plants, you can make a light garden.

A light garden consists of shelves and fluorescent lights to illuminate them. You can purchase a light garden in a garden center or from a garden tool catalog. It can be set up anyplace in your house—even in a basement, if it is warm enough.

WATERING

All houseplants need regular watering. Make sure they are in pots that have a hole in the bottom, so that excess water can run out. Set the pots onto saucers. Every time you water, give the plant enough that some water drains out the bottom, then empty the saucer. Plants in unglazed clay pots need to be watered more frequently than plants in plastic or another waterproof material.

Water quality is important. Rainwater is the best you can use, because it is soft, which means that it is relatively free of minerals. Also, it does not contain the tap water additives chlorine and fluoride, both of which can harm houseplants over time. Tap water varies in quality, depending on its source and the way it is treated. Some tap water is almost free of minerals; other tap water is very hard, meaning that it contains a lot of minerals. These minerals can eventually leave a salty crust on the soil surface and on the pot. If the water is alkaline, it will affect the nutrient balance in the soil, which can lead to unhealthy plants. The ideal water for most houseplants is slightly acidic, as rainwater generally is.

Because of their tropical origin, most indoor plants can be harmed by cold, so water should be at room temperature. Fill a pitcher with tap water and let it stand overnight. Any chlorine will dissipate, and the water will reach room temperature.

Most houseplants should be watered when the soil just under the surface is still slightly moist. The soil will feel cool to the touch but won't make your finger wet or muddy. If the soil is dusty dry, you have waited too long.

However, some houseplants, such as ficus trees, should always be kept moist.

Succulents and cactus have low water requirements. Cactus should be allowed to dry out between waterings. Succulents can be watered regularly in summer, when they will grow fairly rapidly, but in winter, too much water will kill them.

Because fertilizers and mineral salts from water slowly accumulate in potting soil, at least once a year the plants should be placed in the shower, or outdoors in a warm rain, allowing water to run freely through the soil for several minutes. This will dissolve, or leach, excess salts from the soil. Plants that are watered with hard water should be leached more often.

Houseplants also catch dust. Rain or warm shower water will wash the leaves. If dust accumulates before the plants need leaching, the leaves can be wiped with a damp cloth.

FERTILIZING

Houseplants need regular fertilizing. In general, the faster the plant grows, the more fertilizer it requires (although you must be careful not to overfertilize).

One way to feed is to add a little fertilizer—either a liquid or a soluble powder—to the water whenever you water the plant, using about one-fourth the dose recommended on the fertilizer package.

The other way to feed plants is with a slow-release fertilizer. These come in the form of pellets or sticks that are placed in the soil to dissolve slowly, over several months. Mark your calendar to remember to replace them.

Both of these methods provide small amounts of nutrients as the plant needs them, without supplying too much fertilizer at once.

REPOTTING

Some plants, especially those in dim locations, can go years without repotting. But if a plant needs daily watering, or if it threatens to topple over, it should be transplanted to a pot one size larger than the current one.

Put a piece of newspaper over the drainage hole in the new pot and cover with enough purchased potting soil so that the base of the stem will be about an inch below the rim. Turn the plant upside down, holding the stem or the entire plant with your hand and knocking the pot to loosen the roots. Place the entire rootball in the new pot and press more potting soil around the roots. Water thoroughly.

FOLIAGE HOUSEPLANTS

Large Palm

Pygmy date palm
Phoenix roebelenii

The pygmy date palm, a small relative of the date palm of biblical history and Old World commerce, has an airy look because of its slender leaflets. It does best in the South, where the humid indoor conditions it appreciates are likely to be found. During the summer, the soil should be constantly wet, though it must drain thoroughly. In winter, the soil surface should be allowed to dry between waterings. The fernlike fronds die if the soil dries out. The pygmy date palm does best in a temperature range of 50° to 70° F and in bright light.

Large Rounded Houseplant

Dumbcane
Dieffenbachia maculata

Named for its poisonous sap, which causes swelling of the tongue, constriction of the throat, and thus speechlessness if you get it in your mouth, dumbcane is a plant to respect. Otherwise it is easy and without offense. It can also be beautiful, with its big, waxy leaves

Pygmy date palm has a light and open texture in spite of its size. It can decorate a patio or deck during the summer, and spend the winter in a bedroom or living room.

Left: Dieffenbachia, a plant of the rain forests, lends a tropical lushness to a temperate interior.
Right: Umbrella-plant is available in this large size, and also in a dwarf, more contained, size.

streaked with cream or white. It can grow 7 feet or taller in several years. Give this plant a place with some sun, near a west or east window. It does best in temperatures no lower than 55° F. It should be watered whenever the surface of the soil is dry.

Large Upright Houseplant

Umbrella-plant
Schefflera actinophylla

Formerly known as *Brassaia,* umbrella-plant has flat circles of waxy green foliage held far out from the strong, woody stem, like umbrella spokes. This Australian plant can grow 6 feet or taller but can be pruned back if it grows too tall. It needs an area that is bright, but it should not have full sun for more than an hour or two a day, preferably an eastern or western exposure. It does best in a temperature range of 55° to 70° F. It should be watered whenever the soil surface is dry.

Large Weeping Houseplant

Weeping fig
Ficus benjamina

This lovely plant easily becomes a small tree when content. Weeping fig suffers when moved, however, so beginners often give up soon after they bring their new plants home. The weeping fig usually loses its leaves when it is moved into your house from the garden shop or nursery. Left undisturbed and fairly dry, however, and it will likely grow a new crop of leaves, especially if it is a small plant. Drafts and too little light can also cause the weeping fig to lose its leaves, but left alone in a place that is sheltered and warm, preferably with an eastern or western exposure, it can eventually dominate a small room, growing taller than 6 feet. It does best in a temperature range of 55° to 75° F.

Weeping fig has a resting period in winter when it should be watered infrequently. In late spring when growth resumes, it can be watered more frequently, but the soil surface should always be allowed to dry between waterings. Overwatering will cause the leaves to dry and brown at the edges.

Weeping fig does not need frequent repotting. It is content when crowded into a pot that looks far too small for its size.

Medium Fern

Boston fern
***Nephrolepis exaltata* 'Bostoniensis'**

An antique among houseplants, Boston fern has long been appreciated for its provision of dependable greenery in unlikely indoor conditions. It was especially popular in England a century ago because it can grow in damp houses that are cool in winter; it does best with winter temperatures above 50° F and summer temperatures around 70° F. The fronds, which can be 2 feet long, drape gracefully over the edges of large pots or hanging baskets, giving any house a bit of the ambience of Victorian England.

Boston fern's only demand is watering whenever the soil surface is dry. Boston fern fronds that turn brown are a symptom of too little water. They can be snipped off at the base and new green growth will follow. In a dry house, Boston fern will do better with occasional misting. In an extremely dry house, a more drought-resistant fern such as *Pteris* would be more appropriate.

Boston fern appreciates shade, or a northern exposure, or a corner close to an east or west window.

Popular for more than a century, Boston ferns graced many of our grandmothers' houses. When a fern becomes too large, remove it from the pot, carve off several clumps from the outside of the rootball, and repot with fresh soil filling the holes. The "pups" make good gifts.

Medium Houseplant for Low Light

Chinese evergreen
Aglaonema crispum

Valued for its tolerance of low-light conditions, Chinese evergreen lights up dark corners with paddle-shaped foliage splashed or washed with silver. It will grow in light conditions as dim as total shade or as bright as a window without direct sun, and will put up with hot or cool temperatures. The short stems give the plant a shrubby appearance, with foliage fanning outward. The cultivar *A.* 'Silver Queen' has the most silver and pale green on the leaves.

Chinese evergreen does best in a shallow pot; it should be in a place free of cold drafts or smoke in the air. In winter, it needs warmth, at least 60° F, and as much humidity as possible. It should be watered whenever the soil surface is dry.

Medium Palm

Bird-of-paradise
Strelitzia reginae

Although this plant is named for its distinctive flower, you may never see the flower indoors. Flowering does not even begin until the plant is 3 to 4 years old, and then only if the plant is big enough and its environment is suitably warm and sunny.

But even without flowers, this is a palm worth growing. It reaches 3 feet tall with lovely blue-green fan-shaped leaves, each leathery leaf on its own fat stalk. Bird-of-paradise does best in partial sun. In summer, it should be watered whenever the soil surface is dry; in winter, it should be watered sparingly and can tolerate fairly cool conditions, as low as 60° F.

Medium Upright Houseplant

Jadeplant
Crassula ovata

Jadeplant, also known as *C. argentea* or *C. portulaceae,* is one of a group of easy plants called succulents. They are desert plants that store their own water supply in their fleshy leaves. Too much water can kill them.

Jadeplant should be watered only about once a month, or if the leaves start to wither. Keeping the plant dry encourages it to develop a strong, woody stem and large leaves. Jadeplant also needs virtually no fertilization. One application a year is enough.

In a bright place, not below 50° F, jadeplant can grow 3 feet tall. It makes a dramatic focal point in a large container, with its attractive, crooked, trunklike stem and waxy leaves whose edges turn reddish in the sun.

Left: Golddust-plant is prized for its unusual speckled variegation, which forms different patterns in different varieties. This is 'Variegata'. Pinch golddust-plant in May to form a dense, bushy plant. Right: Arrowhead-vine can be allowed to climb a pole or trellis, or it can be grown in a bush form, as this one is.

Medium Houseplant With Variegated Leaves

Golddust-plant
***Aucuba japonica* 'Variegata'**

Easily distinguished by dark green leaves with gold speckles, golddust-plant, also called spotted laurel, often shows up in supermarkets, where it may have suffered a certain amount of neglect, a sign of its toughness. It can put up with light or shade and air that is fairly dry or humid. It can stay outdoors in summer in the shade, and in winter can survive temperatures as low as 15° F, making it ideal for a cool hallway or sunporch. The soil surface should be allowed to dry between waterings, and the plant should be fed only when it is actively growing.

Golddust-plant can grow 13 feet tall outdoors, but indoors its height is limited by the size of the pot. As a houseplant it is seldom taller than 5 feet.

Small Houseplant for Bright Light

Mexican snowball
Echeveria elegans

Echeverias are South American succulents, mostly with rosettes of fleshy leaves that grow close to the ground. Many species are cultivated as houseplants.

One of the easiest to find is Mexican snowball, a branching, spreading shrub that can grow a foot tall. Sometimes labeled *E. harmsii,* it has dense rosettes of pointed, hairy leaves that may be slightly translucent and reddish around the margins. It will do well in a spot near a window with some direct sun.

CLIMBING HOUSEPLANTS

Large Climbing Houseplant

Split-leaf philodendron
Philodendron bipennifolium

As this impressive climber grows, the large, leathery, dark green leaves become larger and eventually split into sections, thus the common name split-leaf philodendron. Tied up here and there along the stem, this plant can happily loop its way around windows and the upper edges of walls, growing 30 feet or more. It does best in a warm room, in partial or day-long shade. It should be watered whenever the soil surface is dry.

If growth becomes weak because of too little water or light, split-leaf philodendron can be cut back to the base and will sprout again.

Medium Climbing Houseplant

Arrowhead-vine
***Syngonium podophyllum* hybrids**

Named for the interesting spearhead shapes of white or cream that decorate the green leaves, arrowhead-vine can be grown in a hanging basket or trained as a climber. If you want it to climb, give it a moss stick or rough piece of bark to cling to with its aerial roots.

Arrowhead-vine, also called goosefoot-plant, is almost indestructible. It puts up with darkness and dryness. It does best, however, in filtered light and warm, humid air, and should be watered whenever the soil surface is dry. It can reach about 6 feet tall. The cultivar 'Variegatum' is also called *Nephthytis triphylla.*

Left: Flowering-maples are vining plants outdoors. Pinch them to keep them dense indoors.
Right: Easter cactus shows its relationship to desert cactuses in the vibrant color of its blooms; it differs from them in having no thorns and in its pendulous habit. This is a jungle cactus, growing in the wild on the branches of rainforest trees.

FLOWERING HOUSEPLANTS

Large Flowering Houseplant

Flowering-maple
Abutilon × hybridum

The flowering-maples, which are not true maples at all but mallows, were favorite houseplants a century ago. The plants thrived in the cool, damp houses of Victorian England and New England. The plants have been rediscovered, appreciated anew for their maplelike leaves and the bell-shaped flowers that hang from leaf axils on threadlike stems.

If you buy a budded plant, you may lose most of the flowers—a common circumstance with flowering houseplants. However, if the plant comes into bud in your house, you will likely have a large crop of flowers, which may be orange, red, yellow, rose, apricot, salmon, or white, depending upon the cultivar.

Flowering-maple, sometimes labeled *A. globosum,* prefers a cool place in bright sun. Blooming will be less in any shade. The plant fares best with a winter temperature range of 45° to 55° F and a summer range of 60° to 65° F. The soil should be kept constantly wet during the flowering season but allowed to dry out slightly between waterings during the growing season, when it should be fertilized once a month.

The flowering-maples favored by the Victorians were tall—20 feet or so—and required constant pruning. Now, however, there are improved dwarf cultivars that are much easier to keep, such as red 'Clementine Variegated' and yellow-flowering 'Moonchimes'.

Medium Cactus

Easter cactus
Hatiora gaertneri

Easier to coax into flowering than the Christmas cactus (*Schlumbergera truncata*), the Easter cactus may not bloom exactly on time for the holidays, but it needs little in the way of special care. Do not be misled by the word *cactus,* however. This is not a desert cactus but a jungle cactus that will die without water. When it is dormant in fall, it should be watered whenever the soil dries out completely—about every 2 weeks. This cool, fairly dry rest period encourages flowering. As soon as flower buds form, watering should be more frequent, whenever the soil surface dries out. Easter cactus does best with an east or west exposure. For the best blooms, summer the plant outdoors in a shady spot.

Easter cactus has reclining stems that make it well suited to hanging baskets. The scarlet flowers bloom on the stem tips.

Left: Crown-of-thorns, a semisucculent, grows either indoors or outdoors. As a houseplant, it produces its best blooms when placed in a sunny window. Right: Moth orchids are known as windowsill orchids because, unlike most, they will grow as a houseplant. Each graceful spray of flowers will last for more than a month.

Medium Shrub

Shrimpplant
Justicia brandegeana

Named for the bright pink color and curving shape of the bracts, shrimpplant (formerly called *Beloperone guttata*) is a soft shrub with shiny leaves an inch or longer. The true flowers are small, white, and almost hidden by the bracts, which can remain showy for weeks after flowering is finished. The plant can grow as tall as 3 feet but should be replaced when it becomes leggy. It prefers a warm place in a south, east, or west window, where it can receive some direct sun. Pinching back the top half of the stems every spring encourages bushiness.

Shrimpplant can tolerate winter temperatures as low as 45° F, when it should be watered sparingly. It does best with cool winters and warm summers. From spring through fall, it should be watered whenever the soil surface is dry.

Small Succulent

Crown-of-thorns
Euphorbia milii* var. *splendens

Impressive prickly spines and tiny, bright red flowers from early spring until midsummer characterize crown-of-thorns, also labeled *E. bojeri*. This is a semisucculent, which means it requires little water, especially in winter, when it should be allowed to dry out totally between waterings. The leaves may drop in winter, but new foliage will appear in spring. Crown-of-thorns does best in a southern exposure, though it will survive with less sun. In summer, it should be fed every 2 weeks to encourage flowering, and watered whenever the soil surface is dry.

If you snap off a stem of crown-of-thorns, wash your hands; the milky sap is poisonous. It irritates the skin and can be acutely painful if it comes in contact with your eyes or mouth.

Orchid

Moth orchid
***Phalaenopsis* hybrids**

Orchids are not plants for neglectful gardeners, as a rule, but moth orchids, named for the shape of the flowers, are tough and not fussy—for orchids, that is. They do need a spot that is humid and warm, perhaps a sunporch or a bright but not sunny window where the plant pot can be situated on small stones set in a shallow tray of water. The soil should not be allowed to dry out, and the plant should be fed once a week. When flowering begins, the orchid can be set on a coffee table or other less climatically suited place. Flowering lasts several months, something to be proud of.

INDEX

Page numbers in boldface type indicate principal references; page numbers in italic type indicate references to illustrations and photographs.

U.S. Measure and Metric Measure Conversion Chart

	Formulas for Exact Measures				Rounded Measures for Quick Reference		
	Symbol	When you know:	Multiply by:	To find:			
Mass (weight)	oz	ounces	28.35	grams	1 oz		= 30 g
	lb	pounds	0.45	kilograms	4 oz		= 115 g
	g	grams	0.035	ounces	8 oz		= 225 g
	kg	kilograms	2.2	pounds	16 oz	= 1 lb	= 450 g
					32 oz	= 2 lb	= 900 g
					36 oz	= 2¼ lb	= 1000 g (1 kg)
Volume	pt	pints	0.47	liters	1 c	= 8 oz	= 250 ml
	qt	quarts	0.95	liters	2 c (1 pt)	= 16 oz	= 500 ml
	gal	gallons	3.785	liters	4 c (1 qt)	= 32 oz	= 1 liter
	ml	milliliters	0.034	fluid ounces	4 qt (1 gal)	= 128 oz	= 3¾ liter
Length	in.	inches	2.54	centimeters	⅜ in.	= 1.0 cm	
	ft	feet	30.48	centimeters	1 in.	= 2.5 cm	
	yd	yards	0.9144	meters	2 in.	= 5.0 cm	
	mi	miles	1.609	kilometers	2½ in.	= 6.5 cm	
	km	kilometers	0.621	miles	12 in. (1 ft)	= 30.0 cm	
	m	meters	1.094	yards	1 yd	= 90.0 cm	
	cm	centimeters	0.39	inches	100 ft	= 30.0 m	
					1 mi	= 1.6 km	
Temperature	° F	Fahrenheit	5/9 (after subtracting 32)	Celsius	32° F	= 0° C	
	° C	Celsius	9/5 (then add 32)	Fahrenheit	212° F	= 100° C	
Area	in.^{2}	square inches	6.452	square centimeters	1 in.^{2}	= 6.5 cm^2	
	ft^2	square feet	929.0	square centimeters	1 ft^2	= 930 cm^2	
	yd^2	square yards	8361.0	square centimeters	1 yd^2	= 8360 cm^2	
	a.	acres	0.4047	hectares	1 a.	= 4050 m^2	